# Studies in Computational Intelligence

Volume 1278

The series "Studies in Computational Intelligence" (SCI) publishes new developments and advances in the various areas of computational intelligence—quickly and with a high quality. The intent is to cover the theory, applications, and design methods of computational intelligence, as embedded in the fields of engineering, computer science, physics and life sciences, as well as the methodologies behind them. The series contains monographs, lecture notes and edited volumes in computational intelligence spanning the areas of neural networks, connectionist systems, genetic algorithms, evolutionary computation, artificial intelligence, cellular automata, self-organizing systems, soft computing, fuzzy systems, and hybrid intelligent systems. Of particular value to both the contributors and the readership are the short publication timeframe and the world-wide distribution, which enable both wide and rapid dissemination of research output.

Indexed by SCOPUS, DBLP, WTI AG (Switzerland), zbMATH, SCImago.

All books published in the series are submitted for consideration in Web of Science.

Srija Roy · Manish Kumar Goyal

# Environmental Monitoring with Integrated Earth Observation Data and Machine Learning

Springer

Srija Roy
Department of Civil Engineering
Indian Institute of Technology
Indore, Madhya Pradesh, India

Manish Kumar Goyal
Department of Civil Engineering
Mehta Family School of Sustainability
Indian Institute of Technology
Indore, Madhya Pradesh, India

ISSN 1860-949X ISSN 1860-9503 (electronic)
Studies in Computational Intelligence
ISBN 978-3-032-19934-8 ISBN 978-3-032-19935-5 (eBook)
https://doi.org/10.1007/978-3-032-19935-5



This Springer imprint is published by the registered company Springer Nature Switzerland AG
The registered company address is: Gewerbestrasse 11, 6330 Cham, Switzerland

*To the watchful eyes in the sky, and the breathing Earth beneath, may their dialogue in data inspire understanding, action, and hope for the generations to come.*

# Foreword

The need for accurate, timely, and intelligent environmental monitoring of the Earth has become especially indispensable in this era marked by rapid environmental changes and challenges. From deforestation and droughts to shifting coastlines and urban expansion, the planet speaks through subtle signals captured not by human eyes, but by satellites silently orbiting above. This book, *Environmental Monitoring with Integrated Earth Observation Data and Machine Learning*, offers a compelling synthesis of two transformative forces: the technological reach of Earth observation and the computational power of machine learning. In bringing these together, the authors, Prof. Manish Kumar Goyal and Dr. Srija Roy, present a timely contribution to the scientific community and a powerful resource for practitioners, students, and decision-makers alike. Drawing from years of research and applied expertise, the authors illuminate how earth observation datasets can be translated into actionable environmental monitoring insight. What makes this work stand apart is not just its technical depth, but its clear vision: to enable better environmental stewardship through smarter tools and deeper understanding. This book thus arrives at a moment when climate change, biodiversity loss, and resource depletion are reshaping our relationship with nature. By empowering readers with the knowledge to decode Earth's patterns using advanced tools, it lights a path forward-one where data-driven decisions become the cornerstone of sustainability. It is both a guide and a call to action.

Indore, India

Srija Roy
Manish Kumar Goyal

# Preface

The idea for this book was born out of years of curiosity, fieldwork, and long hours spent decoding satellite imageries and training machine learning models all driven by a single goal: to better understand our environment and the changes and challenges unfolding within it. As researchers working at the intersection of Earth observation and environmental engineering, we have witnessed how access to remotely sensed data and advancements in data science can transform the way we monitor the environment. Yet, we also recognized a persistent gap between data availability, the derivation of meaningful insights, and the transformation of those insights into implementable actions. Thus, this book is our attempt to bridge that gap. Designed for students, researchers, and practitioners alike, it blends foundational concepts of Earth observation and machine learning with demonstrations of practical tools and techniques. We have strived to present complex methods in an accessible manner, supported by real-world examples and case studies drawn from our own research experiences. We hope it not only informs but also inspires future work at the nexus of environmental science, earth observation, and emerging machine learning technologies.

Indore, India

Srija Roy
Manish Kumar Goyal

**Acknowledgments** We express our sincere gratitude to the Ministry of Jal Shakti, Government of India, for funding the *Narmada River Basin Management Studies* under Project No: NRCB: J24037/3/2020/NRCB. This support has been instrumental in enabling the research that forms the foundation of this book. We also acknowledge the collaborative environment and infrastructure provided through this project which allowed us to explore, test, and refine the integration of Earth observation data with machine learning techniques in the context of environmental monitoring.

**Competing Interests** The authors have no competing interests to declare that are relevant to the content of this manuscript.

**Ethics Approval** This study did not involve any human or animal subjects and thus did not require ethical approval.

# About This Book

This book, titled *Environmental Monitoring with Integrated Earth Observation Data and Machine Learning*, presents a comprehensive exploration of how remote sensing technologies, Earth observation datasets, and advanced data analytics using machine learning can be integrated to monitor, assess, and manage environmental systems. It bridges theoretical foundations with practical applications, offering readers insights into Earth observation principles, satellite data processing, and machine learning algorithms tailored for environmental studies. Featuring real-world case studies and implementable workflows, the book serves as a valuable resource for students, researchers, and practitioners working at the intersection of environmental science, geospatial technologies, and data-driven decision-making for environmental monitoring.

# Contents

# About the Authors

**Dr. Srija Roy** is a Research Associate in the Department of Civil Engineering at IIT Indore, working on the *Condition Assessment and Monitoring of the Narmada River Basin* project funded by the Ministry of Jal Shakti. Her research focuses on Remote Sensing, GIS, and Water Resources Engineering, with particular emphasis on the application of Machine Learning and Artificial Intelligence in Earth system and hydrological studies. She holds a B.Tech. in Civil Engineering from MAKAUT, an M.Tech. in Environmental and Water Resources Engineering from NIT Manipur, and a Ph.D. from NIT Jamshedpur. Dr. Roy has published in reputed international journals and has presented her work at leading national and international platforms such as AGU, EGU, and HYDRO. She serves as a peer reviewer for journals like *Scientific Reports*, *Water Resources Management*, *Environmental Earth Sciences*, and *Water and Climate Change*, and recently received the Jal Vigyan Puraskar (2025) from the Indian Society of Hydraulics.

**Prof. Manish Kumar Goyal** is a professor in the Department of Civil Engineering at IIT Indore and holds the prestigious Chair Professorship for BIS Standardization. His research encompasses water resources engineering, GIS, remote sensing, and climate change. He earned his B.Tech. from NIT Warangal and M.Tech. and Ph.D. from IIT Roorkee, with collaborative research at the University of Waterloo, Canada. He pursued postdoctoral fellowships at NTU Singapore and McGill University, Canada. With over 300 publications and 8221 citations, he has an h-index of 55 and i-index of 142. His research contributions have been recognized with the Prof. R. J. Garde Research Award and the Prof. S. N. Gupta Memorial Lecture Award from the Indian Society of Hydraulics. Internationally, he was honored with the ASCE Best Theoretical Oriented Paper Award and has been consistently ranked among Stanford University's Top 2% Scientists (2020–2025). His edited book *Sustainability: Fundamentals and Applications* (Wiley) also received global acclaim. He serves as an associate editor for the Nature *Scientific Reports*, *PLOS Climate*, *ASCE Journal of Hazardous, Toxic, and Radioactive Waste.* His work on critical infrastructure and drought vulnerability has influenced policy frameworks for sustainable development.

# Abbreviations

| | |
|---|---|
| ALOS | Advanced Land Observing Satellite |
| ALOS-2 | Advanced Land Observing Satellite-2 |
| API | Application Programming Interface |
| AppEEARS | Application for Extracting and Exploring Analysis Ready Samples |
| AQI | Air Quality Index |
| ASCII | American Standard Code for Information Interchange |
| ASTER | Advanced Spaceborne Thermal Emission and Reflection Radiometer |
| AW3D30 | ALOS World 3D 30 m |
| AWS | Amazon Web Services |
| BAU | Business-as-Usual (scenario) |
| CartoDEM | Cartosat Digital Elevation Model |
| CEOS | Committee on Earth Observation Satellites |
| CHIRPS | Climate Hazards Group InfraRed Precipitation with Station data |
| CLI | Command-Line Interface |
| COG | Cloud-Optimized GeoTIFF |
| DAAC | Distributed Active Archive Center |
| DEM | Digital Elevation Model |
| DIAS | Data and Information Access Services |
| DT | Decision Tree |
| ECMWF | European Centre for Medium-Range Weather Forecasts |
| EO | Earth Observation |
| EOS | Earth Observing System |
| EOSDIS | Earth Observing System Data and Information System |
| EOSPSO | Earth Observing System Project Science Office |
| ERA5 | Fifth generation ECMWF reanalysis |
| ESA | European Space Agency |
| FABDEM | Fine-resolution Digital Elevation Model with Vegetation Correction |
| FCC | False-Color Composite |
| GCVI | Green Chlorophyll Vegetation Index |

| | |
|---|---|
| GEE | Google Earth Engine |
| GeoTIFF | Geographic Tagged Image File Format |
| gHM | Global Human Modification |
| GIS | Geographic Information System |
| GMTED2010 | Global Multi-resolution Terrain Elevation Data 2010 |
| GNSS | Global Navigation Satellite Systems |
| GPM | Global Precipitation Measurement |
| GRACE | Gravity Recovery and Climate Experiment |
| GTOPO30 | Global 30 arc-second topography |
| HDF | Hierarchical Data Format |
| HDF-EOS | Hierarchical Data Format Earth Observing System |
| IGS | International GNSS Service |
| IMD | India Meteorological Department |
| IMERG | Integrated Multi-satellite Retrievals for GPM |
| INSAT | Indian National Satellite System |
| ISODATA | Iterative Self-Organizing Data Analysis Technique |
| ISRO | Indian Space Research Organisation |
| ITA | Innovative Trend Analysis |
| JAXA | Japan Aerospace Exploration Agency |
| JSON | JavaScript Object Notation |
| LANCE | Land, Atmosphere Near-real-time Capability for EOS |
| LiDAR | Light Detection and Ranging |
| LST | Land Surface Temperature |
| LULC | Land Use and Land Cover |
| MDVI | Modified Difference Vegetation Index |
| MERIS | Medium Resolution Imaging Spectrometer |
| ML | Machine Learning |
| MMK | Modified Mann–Kendall |
| MNDWI | Modified Normalized Difference Water Index |
| MODIS | Moderate Resolution Imaging Spectroradiometer |
| MSI | Multi-Spectral Instrument |
| NAIP | National Agriculture Imagery Program |
| NASA | National Aeronautics and Space Administration |
| NDMI | Normalized Difference Moisture Index |
| NDVI | Normalized Difference Vegetation Index |
| NDWI | Normalized Difference Water Index |
| NetCDF | Network Common Data Form |
| NEXRAD | Next-Generation Radar |
| NIR | Near-Infrared |
| NOAA | National Oceanic and Atmospheric Administration |
| OGC | Open Geospatial Consortium |
| OLI | Operational Land Imager |
| ORCID | Open Researcher and Contributor ID |
| PCA | Principal Component Analysis |
| PGC | Polar Geospatial Center |

| | |
|---|---|
| RCP | Representative Concentration Pathway |
| RF | Random Forest |
| RFR | Random Forest Regressor |
| RINEX | Receiver Independent Exchange (format) |
| SAFE | Standard Archive Format for Europe |
| SAR | Synthetic Aperture Radar |
| SAVI | Soil-Adjusted Vegetation Index |
| SIO | Scripps Institution of Oceanography |
| SMAP | Soil Moisture Active Passive |
| SP3c | Standard Product 3 Orbit Format, version c |
| SPI | Standardized Precipitation Index |
| SR | Simple Ratio |
| SRTM | Shuttle Radar Topography Mission |
| SSP | Shared Socioeconomic Pathway |
| STAC | SpatioTemporal Asset Catalog |
| SWIR | Shortwave Infrared |
| TanDEM-X | TerraSAR-X add-on for Digital Elevation Measurement |
| TIR | Thermal Infrared |
| UCAR | University Corporation for Atmospheric Research |
| USGS | United States Geological Survey |
| VIIRS | Visible Infrared Imaging Radiometer Suite |
| VNIR | Visible and Near-Infrared |
| WHO | World Health Organization |
| WMS | Web Map Service |
| WMTS | Web Map Tile Service |
| WRI | World Resources Institute |

# List of Figures

# List of Tables

# Chapter 1
# Foundations of Earth Observation

## 1.1 Introduction

This underlying concept of observing Earth from space to study environmental processes emerged and evolved from early aerial reconnaissance techniques used during World War II. Consecutively, the launch of the Soviet satellite Sputnik 1 on October 4, 1957, marked the dawn of satellite-based Earth Observation and demonstrated that space technology could play a transformative role in global monitoring. This was followed by the launch of the United States' Explorer 6 (August 1959) which captured the first-ever image of Earth from space and provided a broader and more systematic perspective of our planet (Kuznetsov et al., 2015). Thus, these early missions laid the groundwork for systematic Earth Observation and EO-based environmental monitoring. However, before the advent of these satellite-based Earth Observations, data collection relied on ground-based surveys, aerial photography, and manual observations. These methods were time-consuming, limited in scope, and lacked the comprehensive coverage that EO datasets provide. Monitoring large-scale temporal and spatial environmental changes or assessing global agricultural patterns was also extremely challenging due to the fragmented nature of ground-based data.

A major milestone in EO was thus reached in 1960 with the launch of TIROS-1 (Television Infrared Observation Satellite), the first weather satellite designed to capture real-time cloud cover images. This advancement laid the foundation for continuous atmospheric and climate monitoring and led to improved weather forecasting and early detection of extreme weather events (Sternberg & Schnapf, 1961). Following this, the launch of Landsat-1 in 1972 (initially called the Earth Resources Technology Satellite) marked a significant turning point for environmental monitoring and research. As the first satellite dedicated to systematic and long-term land cover monitoring, it provided continuous, high-resolution imagery (1 arc-second) of forests, agricultural lands, water bodies, and urban expansion.

These long-term environmental monitoring datasets thus became instrumental in tracking deforestation, desertification, wetland loss, and the effects of climate

 
S. Roy and M. K. Goyal, *Environmental Monitoring with Integrated Earth Observation Data and Machine Learning*, Studies in Computational Intelligence 1278,
https://doi.org/10.1007/978-3-032-19935-5_1

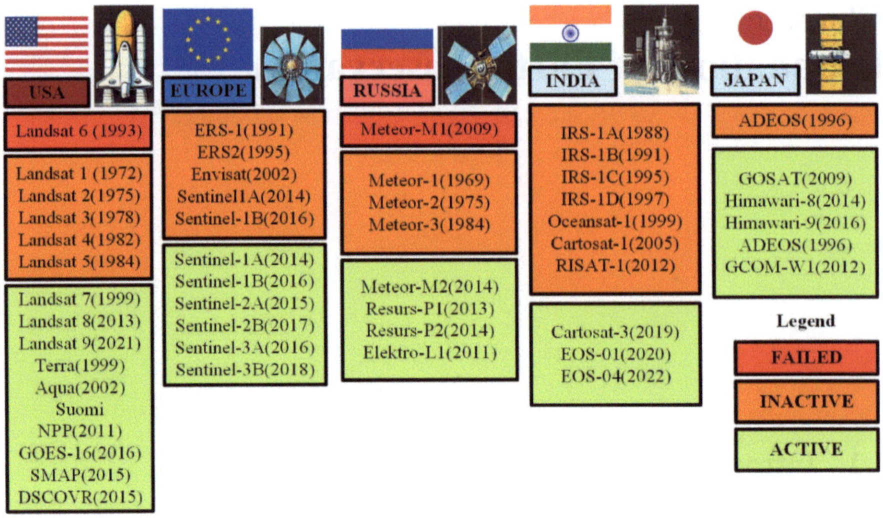


**Fig. 1.1** Country-wise list of major EO missions for environmental monitoring

change on vegetation health. The Landsat program remains one of the most significant Earth Observation initiatives, with decades of continuous data contributing to land use change analysis and ecosystem monitoring (Jain et al., 2025; Rakkasagi et al., 2025; Singh et al., 2024; Williams et al., 2006; Wulder et al., 2022). Similarly, NASA's Nimbus Program introduced a series of satellites that pioneered environmental monitoring of the oceans and atmosphere between 1964 and 1978. These Nimbus satellites also provided early data on sea surface temperatures, ice coverage, and atmospheric composition, significantly improving the understanding of global climate systems. A country-wise list of all major EO missions launched from 1970 to the present is shown in Fig. 1.1. Hence, these early Earth Observation missions set the stage for modern climate change research and environmental impact monitoring (Borfecchia et al., 2010; Denis et al., 2016; Roy, 2024; Rautela et al., 2024a, 2024b; Ustin & Middleton, 2021). Further, in recent times, advancements in sensor technology and data processing have further enhanced the accuracy and applicability of Earth Observation data for environmental monitoring.

## 1.2 Earth Observation Satellites and Systems

EO satellites are designed to collect high-resolution data of the Earth's surface, atmosphere, and oceans. The acquired EO datasets thus obtained play a crucial role in environmental monitoring, climate change research, disaster management, and urban planning. EO satellites also enable efficient analysis of global changes by capturing and transmitting large amounts of environmental data. Advancements in Earth observation systems have also facilitated continuous and near-real-time monitoring of the Earth. Moreover, these EO satellites can be broadly classified based on their mode of data acquisition and sensor technology into active and passive satellites.

### *1.2.1 Classification of EO Satellites Based on Mode of Data Acquisition*

Passive satellites rely on external energy sources, such as the Sun, to capture surface information by measuring reflected or emitted radiation from the Earth's surface. This category includes widely known satellites like NASA and USGS's Landsat series, ESA's Sentinel-2 mission, and ISRO's Resourcesat series. These EO satellites are instrumental in monitoring vegetation, land cover, and climate variability. On the contrary, active satellites generate their own signals and analyze their interaction with the Earth's surface, thus making them effective for imaging even under cloud cover or nighttime conditions. Examples of active satellites include ESA's Sentinel-1, ISRO's RISAT series, Germany's TerraSAR-X, and JAXA's ALOS-2. These satellites provide critical data for disaster response, land deformation studies, and maritime surveillance (Rautela et al., 2024b; Ulaby et al., 1981, 1982, 1986; Zhou et al., 2021).

### *1.2.2 Classification of EO Satellites Based on Orbital Characteristics*

In addition to classification based on the mode of data acquisition, the EO satellites can also be categorized based on their orbital characteristics into geostationary satellites and polar-orbiting satellites. Geostationary satellites (GEO) (e.g., NOAA's GOES series and India's INSAT) maintain a fixed position relative to Earth's surface by orbiting approximately 35,786 km above the equator. This positioning allows continuous observation of specific areas, thereby making them ideal for weather forecasting and telecommunications (Jiang et al., 2019; Li, 2014; Roy et al., 2025; Roy & Goyal, 2025; Tarpley, 1979).

In contrast, polar-orbiting satellites (e.g., NASA's MODIS, ESA's Sentinel missions, ISRO's Cartosat series, etc.) traverse the Earth from pole to pole at lower altitudes (between 500 and 1300 km). Thus, their orbital altitude and position enable global coverage as Earth rotates beneath them providing consistent, high-resolution imagery of different aspects of the environment (Cunningham et al., 2004; Dixit & Pandey, 2024; Goldberg et al., 2013; Jain et al., 2023; Kogan, 1995; Roy & Chintalacheruvu, 2024b). A comprehensive depiction of the classification of the EO satellites based on their orbital characteristics is shown in Fig. 1.2. Similarly, the dataset obtained using these EO satellites are also further classified into different categories.

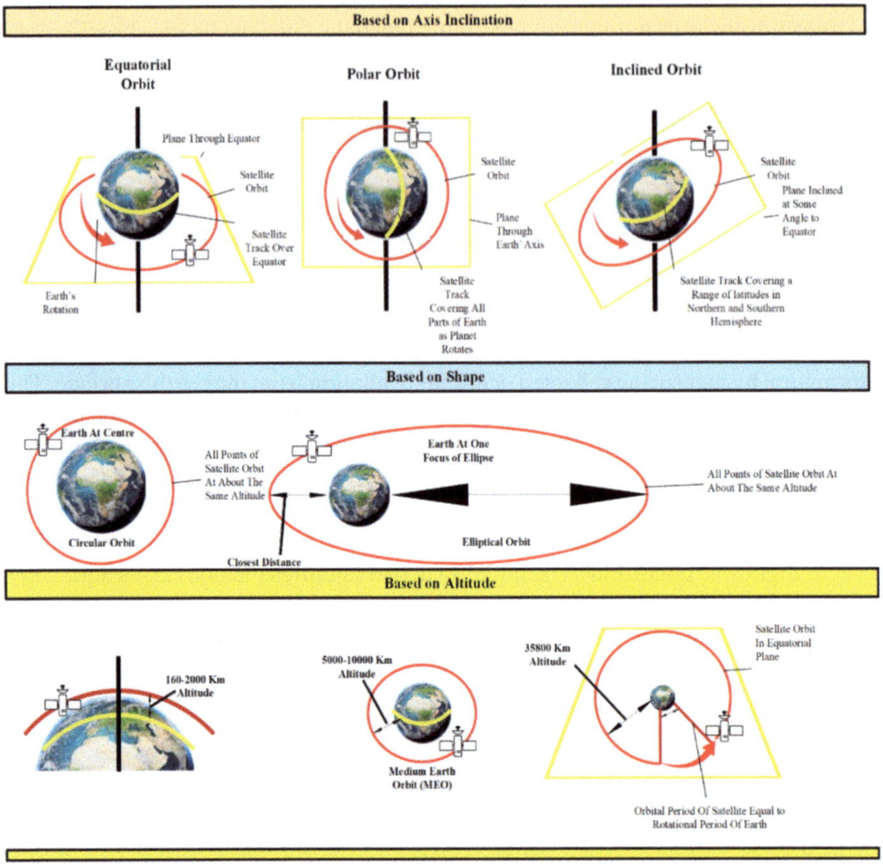


**Fig. 1.2** Classification of EO satellites based on their orbital characteristics

## 1.3 Classification and Characteristics of Earth Observation Datasets

The acquired EO datasets are categorized based on spectral characteristics, spatial resolution, and temporal coverage. This classification is crucial in determining their suitability for specific applications in environmental monitoring.

### *1.3.1 Classification Based on Spectral Characteristics*

EO datasets can be primarily classified into multi-spectral, hyperspectral, and thermal datasets, depending on the number of spectral bands used to capture the raw images.

- **Multi-spectral Dataset**: Multi-spectral EO datasets comprise of a limited number of spectral bands across the visible, near-infrared (NIR), and shortwave infrared (SWIR) regions. These datasets are mostly implemented for vegetation health monitoring, land cover mapping, atmospheric studies, etc. (Stuart et al., 2019).

**Example:** Landsat (OLI), Sentinel-2 (MSI), and MODIS.

- **Hyperspectral Dataset**: Hyperspectral datasets provide multiple contiguous spectral bands enabling the precise material identification for mineral exploration, crop classification, species monitoring, etc. (Dixit et al., 2024).

**Example:** Hyperion, PRISMA, and EnMAP.

- **Thermal Dataset**: Thermal datasets enable capturing of surface temperature variations making them essential for urban heat island (UHI) analysis, drought monitoring, geothermal studies, etc. (Weng, 2009).

**Example:** Landsat (TIRS), ASTER, ECOSTRESS etc.

### *1.3.2 Classification Based on Spatial Resolution*

The spatial resolution of an EO dataset determines the level of detail that can be observed within each pixel of an image (Clark, 1990; Price, 1997).

- **Low Resolution (≥1 km)**: Primarily utilized for global climate modeling, oceanographic studies, large-scale vegetation monitoring, etc.

**Example:** EO dataset obtained from MODIS, AVHRR, GCOM-W, etc.

- **Medium Resolution (10–500 m)**: Widely utilized for urban planning, agriculture, forestry applications, etc. (Roy & Chintalacheruvu, 2023).

**Example:** EO dataset obtained from Landsat, Sentinel-2, Resourcesat, etc.

- **High Resolution (1–10 m)**: High-resolution imageries are utilized for infrastructure mapping, disaster response, precision farming, etc.

**Example:** EO dataset obtained from WorldView, Pleiades, and Cartosat.

- **Very High Resolution (≤1 m)**: These ultra-detailed imageries are primarily utilized for military reconnaissance, cadastral mapping, urban development, etc. (Gong et al., 2019).

**Example:** EO dataset obtained from Maxar's WorldView-3, PlanetScope, GeoEye, etc.

### 1.3.3 Classification Based on Temporal Resolution

EO datasets can also be classified based on their revisit frequency or temporal resolution. This determines how frequently an area on earth is imaged by a particular satellite (Li et al., 2017; Whitcraft et al., 2015).

- **Daily Revisit**: These datasets are primarily utilized for weather monitoring, disaster response, real-time environmental assessments, etc.

**Example:** EO dataset obtained from MODIS, VIIRS, GPM, etc.

- **Weekly to Monthly Revisit**: These datasets are primarily implemented for seasonal crop monitoring, deforestation studies, water resource assessments, etc.

**Example:** Sentinel-2 (5-day revisit), Landsat (16-day revisit), SPOT (daily to biweekly), etc.

- **Yearly or On-Demand Acquisition**: These EO datasets are obtained on custom acquisition basis and caters to specific project needs like infrastructure mapping, archeological explorations, etc.

**Example:** High-resolution commercial satellites, radar-based EO missions, etc.

Following data acquisition, EO datasets undergo preprocessing to ensure their suitability for environmental monitoring applications.

### 1.3.4 Acquisition, Processing, and Utilization of EO Datasets

The raw images acquired from EO satellites are transmitted to ground stations, where they undergo preprocessing to correct distortions caused by sensor limitations, atmospheric interference, and geometric inaccuracies. Common preprocessing techniques comprising of radiometric correction, geometric correction, cloud masking, etc. (Okolie et al., 2022; Panigrahi et al., 2011; Sonka et al., 1993). Following preprocessing, EO data is stored in repositories where it becomes accessible for further processing. The processed datasets are further refined through feature extraction and classification techniques using machine learning (ML) algorithms such as random forest (RF), support vector machines (SVM), and deep learning models to derive significant information for environmental monitoring (Patias et al., 2020; Wilkinson et al., 2024). These processed EO datasets are further utilized across various domains of climate change research, hydrological modeling, disaster response, LULC analysis, biodiversity conservation, etc. (Gerard et al., 2012; Patias et al., 2020).

## 1.4 Earth Observation Datasets for Environmental Monitoring

The rapid advancements in EO missions have significantly enhanced our ability to monitor and analyze environmental changes across various spatial and temporal scales. The availability of high-resolution EO datasets combined with innovations in big data analytics, machine learning (ML), and artificial intelligence (AI) has thus transformed environmental monitoring (Sudmanns, et al., 2020). These technological improvements enable more automated, efficient, and scalable analyses of complex environmental phenomena. A breakthrough in EO-based environmental monitoring came with the development of cloud-based geospatial platforms of Google Earth Engine (GEE), ESA's Sentinel Hub, etc. These platforms facilitate large-scale EO data processing without computational constraints (Roy & Chintalacheruvu, 2024a). Additionally, they provide access to global EO archives and preprocessed datasets, enabling the extraction of valuable insights from multi-temporal datasets for different environmental monitoring applications. Moreover, this cloud-based computational platform also enables the integration of application programming interfaces (APIs) in Python and JavaScript. These APIs can be further utilized to implement ML and AI-driven models for automated classification, anomaly detection, predictive analysis of environmental changes, etc. (Dubey et al., 2021; Goyal et al., 2023). Furthermore, the scalability and real-time processing capabilities of these platforms eliminate the need for high-end computational infrastructure, thus making them widely accessible.

## 1.5 Bridging the Gap: From Datasets to Insights

The transition from raw EO datasets to actionable insights is a crucial step in environmental monitoring and decision-making. While EO satellites generate vast volumes of multi-spectral and hyperspectral datasets, their direct utility depends on effective data processing, analysis, and interpretation (Zhang et al., 2022). Bridging the gap between raw data and meaningful insights requires advanced computational methods, robust data fusion techniques, and a deep understanding of environmental processes to transform observations into valuable intelligence. Thus, the transformation of raw EO datasets into actionable insights requires a systematic multi-stepped approach that integrates multiple methodologies. Data acquisition is followed by preprocessing following which these datasets are analyzed using a combination of ML, statistical modeling, and physical process-based models to extract meaningful patterns and trends (Dixit et al., 2025; Goyal et al., 2023; Gogineni & Chintalacheruvu, 2024; Goyal & Rakkasagi, 2025; Roy & Chintalacheruvu, 2024b; Singh & Kumar, 2024; Zhang et al., 2022). Thus, the choice of analytical techniques depends on the specific environmental application. ML algorithms like random forests (RF), convolutional neural networks (CNNs), support vector machines (SVMs), etc., have become indispensable for land cover classification, deforestation monitoring, disaster risk assessments, etc. (Gogineni & Chintalacheruvu, 2024). Similarly, deep learning models of recurrent neural networks (RNNs), long short-term memory (LSTM) networks, etc., are implemented to monitor and forecast climatic variables, assess ecosystem changes over time (Goyal et al., 2023), etc. Traditional statistical methods of time-series analysis, regression models, principal component analysis (PCA), etc., are also widely used for trend detection in climate variability, hydrological modeling, erosional studies, and air quality assessment (Goyal et al., 2023; Goyal et al., 2024; Roy, 2024; Roy et al., 2024; Singh et al., 2023; Zhang et al., 2022).

Hence, translating EO datasets into actionable insights requires a multi-disciplinary approach that integrates AI-driven analytics, GIS-based spatial modeling, numerical simulations, and multi-source data fusion. These advancements play a crucial role in disaster response, climate adaptation, ecosystem conservation, and overall environmental monitoring.

## References

Borfecchia, F., Pollino, M., De Cecco, L., Lugari, A., Martini, S., La Porta, L., & Pascale, C. (2010). Active and passive remote sensing for supporting the evaluation of urban seismic vulnerability. *Italian Journal of Remote Sensing, 42*(3), 129–141. https://doi.org/10.5721/ItJRS20104239

Clark, R. N., King, T. V., Klejwa, M., Swayze, G. A., & Vergo, N. (1990). High spectral resolution reflectance spectroscopy of minerals. *Journal of Geophysical Research: Solid Earth, 95*(B8), 12653–12680. https://doi.org/10.1029/JB095iB08p12653

Cunningham, J. D., Bloom, H. J., & Nelson, C. S. (2004, September). National polar-orbiting operational environmental satellite system: Meeting future user needs for earth system observations. In *Weather and environmental satellites* (Vol. 5549, pp. 1–10). SPIE. https://doi.org/10.1117/12.561416

Denis, G., de Boissezon, H., Hosford, S., Pasco, X., Montfort, B., & Ranera, F. (2016). The evolution of Earth observation satellites in Europe and its impact on the performance of emergency response services. *Acta Astronautica, 127*, 619–633. https://doi.org/10.1016/j.actaastro.2016.06.019

Dixit, S., & Pandey, K. K. (2024). Spatiotemporal variability identification and analysis for non-stationary climatic trends for a tropical river basin of India. *Journal of Environmental Management, 365*, Article 121692.

Dixit, S., Pandey, K. K., & Kumar, S. (2025). Enhancing non-stationary analysis of extreme precipitation through a precise extreme event extraction approach (No. EGU25–264). In *Copernicus Meetings*.

Dixit, S., Pandey, K. K., & Shukla, D. (2024). Decadal hydrological impact assessment of evolving land use and land cover in an Indian River Basin: A multi-model approach. *Journal of Water and Climate Change, 15*(9), 4418–4433.

Dubey, S., Gupta, H., Goyal, M. K., & Joshi, N. (2021). Evaluation of precipitation datasets available on Google Earth Engine over India. *International Journal of Climatology, 41*(10), 4844–4863. https://doi.org/10.1002/joc.7110

Gerard, F., Blank, L., Bunce, R. G. H., Carmel, Y., Caudullo, G., Clerici, N., & Weissteiner, C. J. (2012). Assessing the role of EO in biodiversity monitoring: Options for integrating in-situ observations with EO within the context of the EBONE concept. *Ecological Indicators, 14*(1), 22–31. https://doi.org/10.1016/j.ecolind.2011.06.031

Gogineni, A., & Chintalacheruvu, M. R. (2024). Assessing temporal snow cover variation in the Sutlej River basin using Google Earth Engine and machine learning models. *Earth Science Informatics, 17*(1), 455–473. https://doi.org/10.1007/s12145-024-01127-6

Goldberg, M. D., Kilcoyne, H., Cikanek, H., & Mehta, A. (2013). Joint Polar Satellite System: The United States next generation civilian polar-orbiting environmental satellite system. *Journal of Geophysical Research: Atmospheres, 118*(24), 13463–13475. https://doi.org/10.1002/2013JD020389

Goyal, M. K., & Rakkasagi, S. (2025). *Technology–policy interface for combating forest loss and fires*. Springer Nature.

Goyal, M. K., Rakkasagi, S., Shaga, S., Zhang, T. C., Surampalli, R. Y., & Dubey, S. (2023). Spatiotemporal-based automated inundation mapping of Ramsar wetlands using Google Earth Engine. *Scientific Reports, 13*, 17324. https://doi.org/10.1038/s41598-023-44502-2

Goyal, M. K., Rakkasagi, S., Surampalli, R. Y., Zhang, T. C., Erumalla, S., Gupta, A., & U-tapao, C. (2024). Enhancing sustainable development through Spatiotemporal analysis of Ramsar wetland sites in South Asia. *Technology in Society, 79*, Article 102723.

Jain, V., Kumar, S., & Goyal, M. K. (2025). Relationship between daily precipitation extremes and temperature in changing climate across smart cities of Central India. *Journal of Environmental Management, 380*, Article 125036.

Jain, V., Rautela, K. S., & Goyal, M. K. (2023). Ecological restoration: An overview of science and policy regime. *Ecosystem Restoration: Towards Sustainability and Resilient Development*, 1–27.

Jiang, H., Lu, N., Qin, J., Tang, W., & Yao, L. (2019). A deep learning algorithm to estimate hourly global solar radiation from geostationary satellite data. *Renewable and Sustainable Energy Reviews, 114*, Article 109327. https://doi.org/10.1016/j.rser.2019.109327

Justice, C. O., Giglio, L., Korontzi, S., Owens, J., Morisette, J. T., Roy, D., & Kaufman, Y. (2002a). The MODIS fire products. *Remote Sensing of Environment, 83*(1–2), 244–262. https://doi.org/10.1016/S0034-4257(02)00076-7

Justice, C. O., Townshend, J. R. G., Vermote, E. F., Masuoka, E., Wolfe, R. E., Saleous, N., & Morisette, J. T. (2002b). An overview of MODIS land data processing and product status. *Remote Sensing of Environment, 83*(1–2), 3–15. https://doi.org/10.1016/S0034-4257(02)00084-6

Kogan, F. N. (1995). Droughts of the late 1980s in the United States as derived from NOAA polar-orbiting satellite data. *Bulletin of the American Meteorological Society, 76*(5), 655–668. https://doi.org/10.1175/1520-0477(1995)076%3c0655:DOTLIT%3e2.0.CO;2

Kuznetsov, V. D., Sinelnikov, V. M., & Alpert, S. N. (2015). Yakov Alpert: Sputnik-1 and the first satellite ionospheric experiment. *Advances in Space Research, 55*(12), 2833–2839. https://doi.org/10.1016/j.asr.2015.02.022

Li, H. (2014). *Geostationary satellites collocation* (pp. 10–98). Springer. https://doi.org/10.1007/978-3-642-45129-3

Li, J., & Roy, D. P. (2017). A global analysis of Sentinel-2A, Sentinel-2B and Landsat-8 data revisit intervals and implications for terrestrial monitoring. *Remote Sensing, 9*(9), 902. https://doi.org/10.3390/rs9090902

Okolie, C. J., & Smit, J. L. (2022). A systematic review and meta-analysis of digital elevation model (DEM) fusion: Pre-processing, methods and applications. *ISPRS Journal of Photogrammetry and Remote Sensing, 188*, 1–29. https://doi.org/10.1016/j.isprsjprs.2022.03.017

Panigrahi, N., Mohan, B. K., & Athithan, G. (2011). Pre-processing algorithm for rectification of geometric distortions in satellite images. *Defence Science Journal, 61*(2), 174–180. https://doi.org/10.14429/dsj.61.333

Patias, P., Mallinis, G., Tsioukas, V., Georgiadis, C., Kaimaris, D., Tassopoulou, M., & Riffler, M. (2020). Earth observations as a tool for detecting and monitoring potential environmental violations and policy implementation. *The International Archives of the Photogrammetry, Remote Sensing and Spatial Information Sciences, 43*, 1491–1496. https://doi.org/10.5194/isprs-archives-XLIII-B3-2020-1491-2020

Pérez-Sánchez, J., Senent-Aparicio, J., Martínez Santa-María, C., & López-Ballesteros, A. (2020). Assessment of ecological and hydro-geomorphological alterations under climate change using SWAT and IAHRIS in the Eo River in Northern Spain. *Water, 12*(6), 1745. https://doi.org/10.3390/w12061745

Price, J. C. (1997). Spectral band selection for visible–near infrared remote sensing: Spectral-spatial resolution tradeoffs. *IEEE Transactions on Geoscience and Remote Sensing, 35*(5), 1277–1285. https://doi.org/10.1109/36.628798

Rakkasagi, S., & Goyal, M. K. (2025). Are rapid onset drying events escalating forest fires across India's ecoregions? *Environmental Research Letters.*

Rakkasagi, S., Edris, S. G., & Goyal, M. K. (2025). A socio-technical framework for assessing flash drought risks to key biodiversity areas: A scientists' warning. *Technology in Society, 82*, Article 102893.

Rautela, K. S., Singh, S., & Goyal, M. K. (2024a). Characterizing the spatio-temporal distribution, detection, and prediction of aerosol atmospheric rivers on a global scale. *Journal of Environmental Management, 351*, Article 119675.

Rautela, K. S., Singh, S., & Goyal, M. K. (2024b). Resilience to air pollution: A novel approach for detecting and predicting aerosol atmospheric rivers within Earth system boundaries. *Earth Systems and Environment*, 1–23.

Roy, S. (2024). Decoding the declining mangrove ecosystems: An analysis of carbon dynamics, salinity changes, and deforestation using remote sensing and machine learning. In *AGU24.*

Roy, S., & Chintalacheruvu, M. R. (2023). Assessing long-term changes in the lower Godavari River Basin using geo-spatial technology: An analysis of river bank erosion and accretion. *Sustainable Lifestyle Approaches Towards Environment and Climate Change, EIACP (PC-RP) Special*, 36–49.

Roy, S., & Chintalacheruvu, M. R. (2024a). Google Earth Engine-based morphometric parameter evaluation and comparative analysis of soil erosion susceptibility using statistical and machine learning algorithms in large river basins. *Earth Science Informatics, 17*(1), 75–97. https://doi.org/10.1007/s12145-023-01060-4

Roy, S., & Chintalacheruvu, M. R. (2024b). Enhanced morphometric analysis for soil erosion susceptibility mapping in the Godavari river basin, India: Leveraging Google Earth Engine and principal component analysis. *ISH Journal of Hydraulic Engineering, 30*(2), 228–244.

Roy, S., & Goyal, M. (2025). Climate variability and extremes in Indian River Basins: Trends, regional disparities, and urban risk assessment (No. EGU25-1114). In *Copernicus Meetings.*

Roy, S., Rao, C. M., & Abioui, M. (2024). Evaluation of non-stationary spatial relationship between meteorological-environmental parameters and PM2.5. *Advances in Space Research, 73*(8), 4106–4124.

Roy, S., Uddin, M. G., Abdelrahman, K., Fnais, M. S., & Abioui, M. (2025). Assessing the impact of digital elevation model resolution on hypsometric analysis in large River Basins (India): A non-parametric statistical approach. *Earth Science Informatics, 18*(1), 1–21.

Singh, S., & Kumar, A. (2024). Understanding the intricacies of rainfall dynamics using entropy measures. *Journal of Water and Climate Change, 15*(9), 4817–4839.

Singh, S., Kumar, D., Kumar, A., & Kuriqi, A. (2023). Entropy-based assessment of climate dynamics with varying elevations for hilly areas of Uttarakhand, India. *Sustainable Water Resources Management, 9*(4), 130.

Singh, S., Kumar, D., Vishwakarma, D. K., Kumar, R., & Kushwaha, N. L. (2024). Seasonal rainfall pattern using coupled neural network-wavelet technique of southern Uttarakhand, India. *Theoretical and Applied Climatology, 155*(6), 5185–5201.

Sonka, M., Hlavac, V., & Boyle, R. (1993). Image pre-processing. In *Image processing, analysis and machine vision* (pp. 56–111). Chapman and Hall.

Sternberg, S., & Schnapf, A. (1961). Performance and evaluation of satellites TIROS I and TIROS II. *ARS Journal, 31*(11), 1495–1505. https://doi.org/10.2514/8.5635

Stuart, M. B., McGonigle, A. J., & Willmott, J. R. (2019). Hyperspectral imaging in environmental monitoring: A review of recent developments and technological advances in compact field deployable systems. *Sensors, 19*(14), 3071. https://doi.org/10.3390/s19143071

Sudmanns, M., Tiede, D., Lang, S., Bergstedt, H., Trost, G., Augustin, H., & Blaschke, T. (2020). Big Earth data: Disruptive changes in Earth observation data management and analysis? *International Journal of Digital Earth, 13*(7), 832–850. https://doi.org/10.1080/17538947.2019.1585976

Tarpley, J. D. (1979). Estimating incident solar radiation at the surface from geostationary satellite data. *Journal of Applied Meteorology and Climatology, 18*(9), 1172–1181. https://doi.org/10.1175/1520-0450(1979)018%3c1172:EISRAT%3e2.0.CO;2

Ulaby, F. T., Moore, R. K., & Fung, A. K. (1981). *Microwave remote sensing: Active and passive. Volume I: Microwave remote sensing fundamentals and radiometry*. Artech House.

Ulaby, F. T., Moore, R. K., & Fung, A. K. (1982). *Microwave remote sensing: Active and passive. Volume II: Radar remote sensing and surface scattering and emission theory*. Artech House.

Ulaby, F. T., Moore, R. K., & Fung, A. K. (1986). *Microwave remote sensing: Active and passive. Volume III: From theory to applications*. Artech House.

Ustin, S. L., & Middleton, E. M. (2021). Current and near-term advances in Earth observation for ecological applications. *Ecological Processes, 10*, Article 1. https://doi.org/10.1186/s13717-020-00267-x

Weng, Q. (2009). Thermal infrared remote sensing for urban climate and environmental studies: Methods, applications, and trends. *ISPRS Journal of Photogrammetry and Remote Sensing, 64*(4), 335–344. https://doi.org/10.1016/j.isprsjprs.2009.03.007

Whitcraft, A. K., Becker-Reshef, I., Killough, B. D., & Justice, C. O. (2015). Meeting Earth observation requirements for global agricultural monitoring: An evaluation of the revisit capabilities of current and planned moderate resolution optical Earth observing missions. *Remote Sensing, 7*(2), 1482–1503. https://doi.org/10.3390/rs70201482

Wilkinson, R., Mleczko, M. M., Brewin, R. J. W., Gaston, K. J., Mueller, M., Shutler, J. D., & Anderson, K. (2024). Environmental impacts of Earth observation data in the constellation and cloud computing era. *Science of the Total Environment, 909*, Article 168584. https://doi.org/10.1016/j.scitotenv.2023.168584

Williams, D. L., Goward, S., & Arvidson, T. (2006). Landsat. *Photogrammetric Engineering & Remote Sensing, 72*(10), 1171–1178.

Wulder, M. A., Roy, D. P., Radeloff, V. C., Loveland, T. R., Anderson, M. C., Johnson, D. M., & Cook, B. D. (2022). Fifty years of Landsat science and impacts. *Remote Sensing of Environment, 280*, Article 113195. https://doi.org/10.1016/j.rse.2022.113195

Zhang, B., Wu, Y., Zhao, B., Chanussot, J., Hong, D., Yao, J., & Gao, L. (2022). Progress and challenges in intelligent remote sensing satellite systems. *IEEE Journal of Selected Topics in Applied Earth Observations and Remote Sensing, 15*, 1814–1822. https://doi.org/10.1109/JSTARS.2022.3153065

Zhou, Y., Flynn, K. C., Gowda, P. H., Wagle, P., Ma, S., Kakani, V. G., & Steiner, J. L. (2021). The potential of active and passive remote sensing to detect frequent harvesting of alfalfa. *International Journal of Applied Earth Observation and Geoinformation, 104*, Article 102539. https://doi.org/10.1016/j.jag.2021.102539

# Chapter 2
# Earth Observation Data Management and Analytical Tools

## 2.1 Introduction

The integration of EO datasets with in-situ measurements requires robust frameworks and advanced data management methodologies. Therefore, once the EO dataset is acquired, it must undergo rigorous preprocessing followed by advanced analytical and processing workflows (Okolie et al., 2022; Sonka et al., 2013). Thus, structured storage, version control, and metadata tagging are essential aspects of EO dataset management and storage that ultimately ensues their long-term accessibility and usability (Sudmanns et al., 2020).

## 2.2 Earth Observation Dataset Management

The advancements in EO satellite and sensor technologies have enabled analysis of minute details of the earth and its environment. This, however, has resulted in significant increase in volume and variety of EO datasets, thereby making data management more efficient by utilizing the available tools of data formatting, creating data basics glossary, availing artificial intelligence (AI) and cloud computing, and assigning data processing levels. Thus, the EO datasets following acquisition follow a rigorous workflow and undergo the required preprocessing as a part of the EO data management and processing workflow. The different components of Earth Observation datasets are shown in Fig. 2.1. Thus, these complex and unique EO datasets are arranged and tagged in correct data formatting followed by their data cataloging, and updating data records followed by the preprocessing, visualization, processing, and analysis for the extraction of relevant information about the earth and its environment.



S. Roy and M. K. Goyal, *Environmental Monitoring with Integrated Earth Observation Data and Machine Learning*, Studies in Computational Intelligence 1278,
https://doi.org/10.1007/978-3-032-19935-5_2

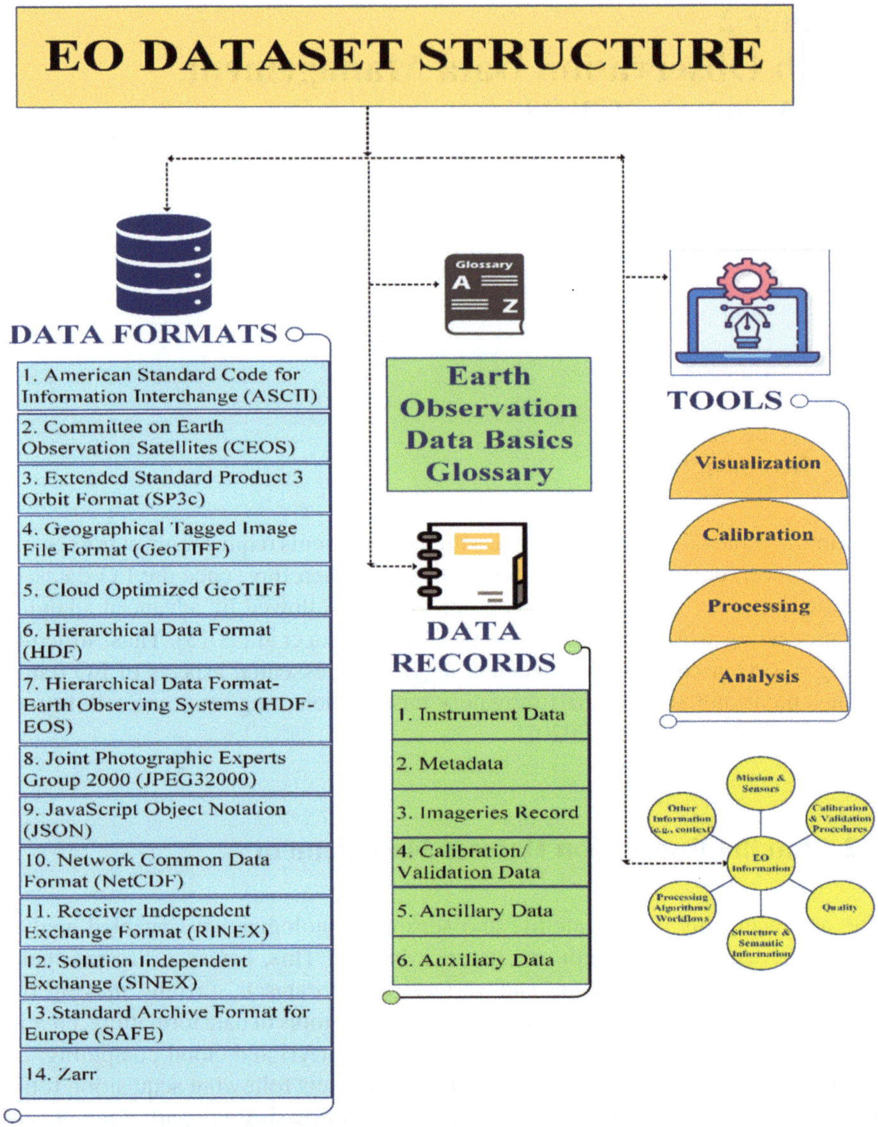


**Fig. 2.1** Constituting components of Earth Observation datasets

## 2.2.1 *Earth Observation Dataset Preprocessing*

Preprocessing EO datasets is the first and most crucial step required before the actual processing and utilization of EO data for environmental monitoring. This initial step optimizes the quality, and usability ensures their standardization and long-term usability. The following are the key steps involved:

**Radiometric Correction**: This preprocessing step enables the adjustment of sensor-based errors along with the maintenance of consistency in pixel brightness by correcting the differences between sensor noise and illumination. This is a crucial pre-requisite for EO datasets and is especially important for long-term temporal studies of land use and land cover change, vegetation mapping, etc. (Goyal et al., 2024; Jain et al., 2023; Paolini et al., 2006; Rakkasagi et al., 2025).

**Geometric Correction**: This preprocessing step ensures the proper alignment of the images to a standard coordinate system by removing spatial distortions caused by factors like satellite motion, terrain relief, sensor tilt, etc. Thus, especially for terrain characteristics studies like geomorphological studies, flood mapping, etc., geometric correction is extremely important (Bannari et al., 1995; Li et al., 2021).

**Atmospheric Correction**: This preprocessing step eliminates atmospheric interference like scattering and absorption and improves the overall reflectance accuracy of the images, thereby enhancing the utilizability of EO datasets for air and atmospheric studies (Doxani et al., 2023; Goyal et al., 2024; Mahiny & Turner, 2007; Rautela et al., 2025).

**Cloud and Shadow Masking**: Radiometric, geometric, and atmospheric preprocessing steps are followed by the steps of cloud and shadow cover removal and masking. This step ensures efficient detection and removal of cloud and shadow covers to ensure the rectification of any obscure land surface features in optical EO data (Zhu & Woodcock, 2012; Molthan et al., 2015).

**Resampling and Subsetting**: This final preprocessing step of resampling and subsetting ensures that the pixel sizes of the EO datasets are adjusted and resampled as per user requirements followed by the extraction of the EO datasets for specific regions of interest.

Thus, these preprocessing steps along with other processing enable these EO datasets to be processed at various levels ranging from level 0 to level 4 EO datasets. Level 0 EO datasets are simply the datasets in their raw format at the similar resolution as at its time of acquisition. Since all the Earth Observation system (EOS) instruments and the subsequent datasets derived from them have level 1 standard data product (SDP), most of the data products of Earth Observing System Data and Information System (EOSDIS) are available at level 2 or level 3 SDPs (LTDP Working Group, 2013). Moreover, the datasets that pertain to interdisciplinary science investigations also generate level 4 EO datasets. This segregated data processing levels, and their respective significance enables efficient EO datasets management and is critical for ensuring inter-comparability between datasets from different missions and sensors.

### 2.2.2 Earth Observation Data Processing Levels

EO data products are categorized into standardized data processing levels ranging from level 0 (raw data) to level 4 (model-derived or assimilated products). These processing levels reflect the degree of transformation applied to the raw satellite signals, with each level suited for different analytical needs and application domains. The guidelines for generating these standard data products (SDPs) are consistently maintained by authoritative bodies such as the Earth Observing System Project Science Office (EOSPSO), NASA, and its network of Distributed Active Archive Centers (DAACs), which ensure uniformity and interoperability across EO datasets (Earthdata, 2025).

A summary of these data levels and their respective descriptions is given in Table 2.1, which outlines the characteristics and appropriate use cases for each level of EO data. The choice of which data level to use is often determined by project-specific requirements, balancing the trade-off between data richness and ease of use. For example, lower-level datasets (Levels 0–1), which are minimally processed and retain raw or radiometrically corrected sensor readings, are frequently used in foundational geophysical assessments such as soil properties mapping (Roy et al., 2024; Shangguan et al., 2014) or topographic analysis (Eidenshink & Faundeen, 1994). In contrast, higher-level datasets (Levels 2–4) which include geometrically corrected, atmospherically corrected, or even model-assimilated products are utilized for more complex interdisciplinary studies, such as evaluating the impact of socio-economic factors on land use, assessing human health exposure, or quantifying human-induced environmental changes (Hu et al., 2017; Kennedy et al., 2019). Thus, understanding this EO data level hierarchy is crucial as it ensures methodological accuracy and relevance. It also enables data storage and computational resources optimization. The awareness of which in turn supports better selection and use of datasets for EO-based environmental monitoring.

### 2.2.3 Earth Observation Data Storage and Management

The storage and management of EO datasets have become increasingly complex driven by the exponential growth in both dataset volume and dimensionality. As the scope and scale of EO applications continue to expand, the efficient organization and maintenance of these datasets have become essential for ensuring their long-term interpretability, accessibility, and usability across scientific, environmental, and policy-driven domains. To facilitate these goals, EO datasets are typically stored in standardized digital formats that encapsulate not only raw or processed data but also comprehensive metadata. These metadata provide crucial information about the dataset structure, spatial–temporal coverage, and variable characteristics. This in turn enables efficient indexing, discovery, and usage of the acquired EO datasets. Furthermore, these standardized formats play a pivotal role in ensuring interoperability, consistency, and seamless data sharing across different platforms and institutions.

**Table 2.1** Summary of Earth Observation data levels and their key characteristics

| SI. No. | EO data level | Dataset type | Dataset resolution | Processing status | Annotations |
|---|---|---|---|---|---|
| 1 | Level 0 | Refined instrument and payload data | Full | Unprocessed | 1. Synchronization frames removed<br>2. Communication headers removed<br>3. Duplicate data removed |
| 2 | Level 1A | Reconstructed instrument data | Full | Unprocessed | 1. Includes ancillary information:<br>2. Radiometric calibration coefficients<br>3. Geometric calibration coefficients<br>4. Georeferencing parameters (e.g., platform ephemeris) |
| 3 | Level 1B | Calibrated instrument data in physical units (e.g., radiance) | Full | Calibrated | 1. Derived from L1A<br>2. Converted to instrument units (e.g., radiance, reflectance)<br>3. Not available for all instruments |
| 4 | Level 1C | Spectrally annotated L1B data | Full | Annotated | 1. Adds variables describing spectral characteristics<br>2. Indicates whether channels are original or synthesized from L1B |
| 5 | Level 2 | Geophysical variables derived from L1 source data | Same as source | Derived | 1. Derived at same resolution/location as L1<br>2. Includes variables like surface temperature, vegetation index |
| 6 | Level 2A | Surface metrics from geolocated data | Variable | Geophysical | 1. Includes ground elevation<br>2. Relative height metrics<br>3. Waveform-derived surface characteristics |
| 7 | Level 2B | Instrument-unit representation of L2A data | Variable | Calibrated | 1. L2A data processed to instrument units<br>2. Not always available for every instrument |
| 8 | Level 3 | Space–time gridded geophysical variables | Gridded | Spatially aggregated | 1. Uniform grid scale<br>2. Enhanced completeness and consistency |

(continued)

**Table 2.1** (continued)

| SI. No. | EO data level | Dataset type | Dataset resolution | Processing status | Annotations |
|---|---|---|---|---|---|
| 9 | Level 3A | Periodic summaries of L2 products | Gridded (temporal) | Temporally aggregated | 1. Weekly, 10-day, or monthly summaries<br>2. Useful for trends and seasonal studies |
| 10 | Level 4 | Model output or multi-source analysis | Variable | Analyzed | 1. Includes model simulations<br>2. Integrates multiple datasets or derived variables |

Among the most widely used formats are the Network Common Data Form (NetCDF), Hierarchical Data Format (HDF), Geographic Tagged Image File Format (GeoTIFF), JavaScript Object Notation (JSON), and ASCII (Fig. 2.2). Notably, NetCDF is particularly suitable for storing array-oriented geophysical variables and is extensively used in fields of meteorology and climatology. It supports platform-independent, self-describing data structures with multiple dimensions and rich metadata, as observed in datasets like NOAA's Global Forecast System and NASA's MODIS products (Thaler, 2016; Zhuk, 2024). Similarly, HDF is optimized for handling large, heterogeneous EO datasets due to its flexible internal structure. This makes it ideal for storing complex hierarchical data from satellite missions like EOS-Terra and Aqua (Duane et al., 2000; Folk et al., 1999) (Table 2.2).

However, as data volumes continue to surge and traditional formats face limitations in network speed and storage costs, there has been a paradigm shift toward cloud-optimized data formats. Emerging formats like the Cloud-Optimized GeoTIFF (COG) and Zarr offer a scalable and efficient alternative. These formats support partial, on-demand access to specific data subsets via HTTP range requests or object storage APIs enabling real-time streaming and analysis without requiring complete downloads. Their implementation in platforms like Microsoft's Planetary Computer and Google Earth Engine has enabled more dynamic and accessible EO data processing (Gowan et al., 2022; Khan & Ramesh, 2024; Lukacz et al., 2024; Roy & Chintalacheruvu, 2024a, 2024b, 2024c). In particular, the Zarr format has proven highly effective for high-resolution, cloud-based atmospheric and geospatial analyses (Gowan et al., 2022). However, despite the technological advantages that these cloud-based solutions provide, their increased usage raises environmental concerns about energy consumption and environmental sustainability. Large-scale cloud data centers can have a significant carbon footprint, which underscores the need for sustainable data management strategies like data compression, cold storage, and the adoption of carbon-aware computing practices (Wilkinson et al., 2024).

Thus, major space agencies like NASA, ESA, ISRO, and JAXA are developing robust archival systems and long-term data stewardship frameworks as a mitigation strategy. Initiatives such as NASA's Earthdata, ESA's Copernicus Open Access Hub,

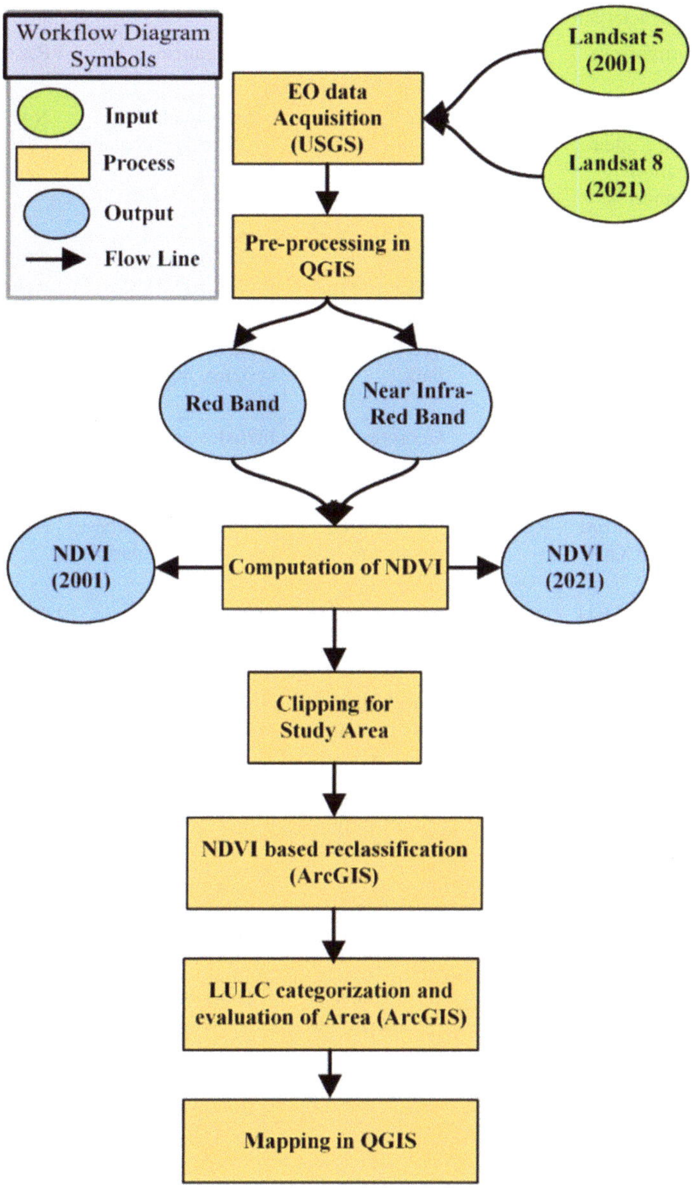


**Fig. 2.2** Methodology flowchart of NDVI-based land categorization of India for 2001 and 2021

**Table 2.2** Characteristics of commonly used EO data formats

| Sl. No | Full form/origin | Type | Primary application | Key features | Recommended by |
|---|---|---|---|---|---|
| 1 | American Standard Code for Information Interchange (ASCII) | Text | Basic Earth science datasets | Human-readable, standard 128-character encoding | NASA ESDIS (2016) |
| 2 | Committee on Earth Observation Satellites (CEOS) | Binary | Radar data (ERS-1, ALOS, etc.) | SAR data split into VOL, LED, IMG, TRL files | CEOS Agencies |
| 3 | Standard Product 3 Orbit Format (SP3c) | Text | Satellite orbit data | Used in GNSS systems, highly precise | IGS |
| 4 | Geographic Tagged Image File Format (GeoTIFF) | Raster | Georeferenced imagery | Includes geolocation, supports map projection | OGC, NASA ESDIS |
| 5 | Cloud-Optimized GeoTIFF (COG) | Raster | Cloud-based geospatial data | Efficient cloud access, backward compatible with GeoTIFF | NASA ESDIS |
| 6 | Hierarchical Data Format (HDF5) | Binary | Large, complex datasets | Supports groups, datasets; platform-independent | The HDF Group |
| 7 | HDF for Earth Observing Systems (HDF-EOS) | Binary | NASA EOS satellite data | Geolocation + science data binding | NASA ESDIS (2007) |
| 8 | JPEG 2000 Image Format (JPEG2000) | Raster | High-resolution satellite images | Lossless & lossy compression, better quality than JPEG | JPEG Committee |
| 9 | JavaScript Object Notation (JSON) | Text | Metadata, configuration, web APIs | Lightweight, language-independent | Widespread Use |
| 10 | Network Common Data Form (NetCDF) | Binary | Atmospheric, oceanographic data | Self-describing, scalable, portable | UCAR/Unidata |
| 11 | Receiver Independent Exchange Format (RINEX) | Text | GNSS raw data | Version 2 for GPS, Version 3 for multi-GNSS | IGS |
| 12 | Solution Independent Exchange Format (SINEX) | Text | Station position & Earth orientation | Official IGS combination product | IGS |
| 13 | Standard Archive Format for Europe (SAFE) | Folder Structure | Sentinel satellite data | XML manifest + metadata/ measurement subfolders | ESA |

and ISRO's Bhuvan Portal exemplify global efforts toward open access, data standardization, and long-term preservation of EO archives (Albani & Giaretta, 2009). Thus, these platforms offer accessible and structured EO datasets. Yet, their potential is unlocked only through meaningful use and integration. This in turn demands tools capable of handling complex datasets followed by the generation of significant insights.

## 2.3 Earth Observation Data Integration Tools and Platforms

Advancements in EO data acquisition, storage, and archiving have improved accessibility and interoperability. However, these alone cannot turn large datasets into actionable insights. Real value emerges when EO data are integrated with analytical tools, decision-support systems, and interoperable platforms. Data integration tools and platforms are thus key to modern EO workflows (Gomes et al., 2020). Data integration involves organizing, preprocessing, and combining datasets from various EO sources (e.g., optical, radar, LiDAR) with in-situ, socio-economic, and modeled data to generate context-rich insights for environmental decision-making. This process is supported by a growing ecosystem of tools and platforms, which can be categorized into desktop GIS platforms, cloud-based geospatial services, application programming interface (API)-driven scripting environments, and artificial intelligence (AI)/machine learning (ML)-integrated frameworks (Cheng et al., 2020).

### *2.3.1 Desktop-Based EO Data Integration Tools and Platforms*

Traditional desktop-based remote sensing (RS) and GIS platforms have played a central role in EO data integration. These platforms provide toolkits for visualization, thematic mapping, spatial analysis, and clustering, thereby enabling the integration of data from multiple sources such as optical (Landsat; Roy & Chintalacheruvu, 2023), radar (SAR; Waring et al., 1995), and LiDAR (Hostetler et al., 2018) sensors. Commercial platforms like ArcGIS (Environmental Systems Research Institute, Inc. (ESRI)), Environment for Visualizing Images (ENVI), and ERDAS Imagine offer advanced capabilities for EO data processing and integration. These tools support raster and vector data in varying formats and facilitate functions like georeferencing (Hinge et al., 2018), image classification (Khan et al., 2019), spatial interpolation (Roy et al., 2024), time-series analysis (Gogineni & Chintalacheruvu, 2024), and change detection (Anand & Oinam, 2022). Features like raster calculations, hotspot analysis, and data overlay tools further enable effective monitoring of environmental changes like deforestation and land cover transformation (Hansen et al., 2013). Additionally, semi-automation through model builders in ArcGIS or Interactive Data Language (IDL) scripting in ENVI allows for tailored project design and workflow customization (Allen, 2011; Xue, 2010).

In contrast, open-source platforms like QGIS, SAGA GIS, and GRASS GIS have emerged as robust alternatives to these commercial platforms. These open-source tools also allow users to write scripts and develop custom plugins. Especially tools like QGIS offers strong integration with Python and GDAL, thereby making it suitable for workflows that combine satellite data with field observations, socio-economic data, etc. (Pérez-Peña et al., 2009; Vitalis et al., 2020). However, the real-world utility and operational relevance of these technologies are best demonstrated through practical case studies. Case-based implementations provide deeper insights into how EO datasets and analytical tools are combined in different contexts to address pressing environmental issues. They not only help validate methodological frameworks but also highlight their scalability, replicability, and performance under field conditions.

#### 2.3.1.1 Example Case Studies

The utility of desktop-based EO data tools was demonstrated through an NDVI-based land classification case study (Roy & Chintalacheruvu, 2024a, 2024b, 2024c). The study integrated preprocessing, classification, and visualization using the EO data integration tools of QGIS and ArcGIS. Landsat 5 (2001) and Landsat 8 (2021) images covering India were acquired from the USGS EarthExplorer. These raw images were preprocessed in QGIS using the Semi-Automatic Classification Plugin (SCP), which applied radiometric corrections and converted digital numbers to surface reflectance (Congedo et al., 2016; Wulder et al., 2018). To handle the large spatial extent, the images were resampled from 30 to 2,500 m resolution using bilinear interpolation. This resampled data were then used to extract red and NIR bands for NDVI computation, with the relevant band details given in Table 2.3.

Following evaluation of NDVI for the years 2001 and 2021, the reclassification of each of the raster layers for the year 2001 and 2021 was done as per the NDVI-based land category classification criteria (Jain et al., 2025; Roy & Chintalacheruvu, 2024a, 2024b, 2024c) in ArcGIS 10.3 using the reclassification and field calculation toolbox. The final mapping of the NDVI layers was performed in QGIS 3.42.1, and the step-by-step diagram of the methodology followed is shown in Fig. 2.2.

**Table 2.3** Comparative spectral band designations of Landsat 5 and Landsat 8 for NDVI evaluation

| Sl. No | Landsat 5 | Landsat 8 |
|---|---|---|
| Red (R) | Band 3 (0.63–0.69 μm) | Band 4 (0.64–0.67 μm) |
| Near-Infrared (NIR) | Band 4 (0.76–0.90 μm) | Band 5 (0.85–0.88 μm) |
| NDVI | $NDVI_{2001} = \frac{B4-B3}{B4+B3}$ | $NDVI_{2021} = \frac{B5-B4}{B5+B4}$ |

#### 2.3.1.2 NDVI-Based Land Categorization of India

The final mapping and land categorization of India using Normalized Difference Vegetation Index (NDVI) was performed. This enabled the assessment of vegetation condition and land dynamics over a 20-year period (2001–2021). The analysis involved categorizing the entire spatial extent of India into multiple NDVI-based vegetation classes followed by the calculation of the areal extent of each land category. The state-level administrative boundaries of India were acquired from the Survey of India (SOI) and were overlaid on the NDVI raster to provide region-specific insights, thus enabling the extraction of state-wise NDVI statistics of maximum, minimum, mean, and standard deviation for each NDVI-based land category (Roy & Chintalacheruvu, 2024a, 2024b, 2024c). Furthermore, the spatial extent of each NDVI-defined land category across India was visualized and is shown in Fig. 2.3. The result of the analysis reveals that the Northeastern states of Manipur, Meghalaya, Arunachal Pradesh, Nagaland, and Tripura host the largest contiguous areas of reserved and protected forest vegetation. Similar forest patches were also observed in the southern Western Ghats region. Visual comparison between the NDVI land categorization maps of 2001 and 2021 suggests a noticeable expansion in reserved and protected forest areas, along with a decline in grasslands and pasture zones, which in several cases appear to have transitioned into cultivated land.

Quantitative estimates of the analysis also show a net increase in forested area categories and a reduction in wasteland and low-vegetation cover zones. This suggests possible impacts of reforestation policies and changes in land use practices (Tambe et al., 2022). The state-wise tabulated values (mean, max, min, and SD) derived through ArcGIS zonal statistics further highlight these trends at the state level. While the mean annual NDVI for 2001 and 2021 offers valuable insights into land cover changes, a long-term average from 2000 to 2021 was also analyzed. This enables to reduce the impact of inter-annual variability and to better capture overall land category trends across India. However, traditional desktop-based tools pose scalability challenges in computing long-term NDVI statistics across such a large geographic extent. Therefore, the study utilized cloud-based Earth Observation

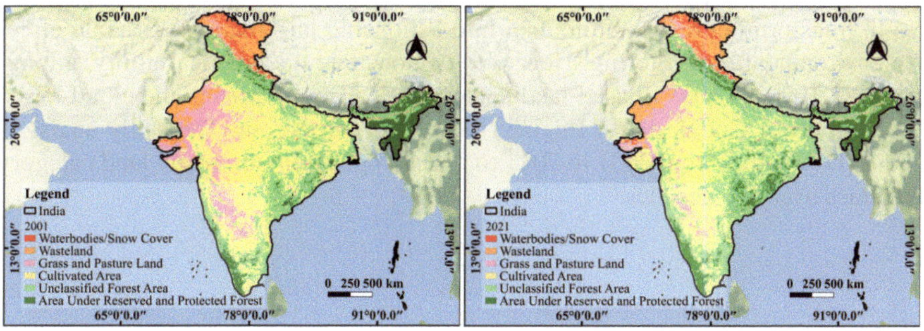


**Fig. 2.3** Spatial mapping of NDVI-based land categorization for 2001 and 2021 respectively

platforms for preprocessing, analysis, and visualization. This shift from standalone desktop-based GIS tools to web-based EO integration environments allowed for efficient data handling, real-time computation, and automated workflows. Thus, this required shift highlights the importance of web-based EO tools not only for high-resolution national-level vegetation monitoring but also for long-term environmental monitoring over greater spatial extents.

### 2.3.2 Web-Based EO Data Integration Tools and Platforms

The increasing volume, complexity, and diversity of EO datasets have necessitated the development of cloud-native platforms that allow scalable, real-time integration and analysis of multi-source EO datasets (Roy & Chintalacheruvu, 2024a, 2024b, 2024c). These web-based computational platforms offer multi-petabyte storage capacity and built-in processing infrastructure, thereby overcoming the limitations of local storage and computational resources. Prominent platforms like Google Earth Engine (GEE), ESA Copernicus DIAS, Sentinel Hub, NASA Earthdata, and AppEEARS further facilitate the seamless integration of multi-source EO datasets, thereby enabling users to access, preprocess, and analyze satellite data from diverse EO missions. These platforms also support API-driven interaction, batch processing, and interactive visualization, making them essential for EO-based environmental monitoring, change detection, and resource management at various spatial and temporal scales. The key features, data sources, and user interfaces of the most used web-based EO dataset integration platforms are given in Table 2.4. This growing ecosystem of platforms not only democratizes EO data access but also lays the foundation for more advanced, intelligent, and automated EO analysis workflows (Roy & Goyal, 2025a; Troumpoukis et al., 2024).

Thus, continuous advances in EO data analysis have also resulted in significant advancements in web-based integration tools to enhance the pace and scalability of environmental insights extraction. With features like multi-source data fusion, large-scale processing, and machine learning support, these platforms are indispensable for researchers, policymakers, and conservationists. The true strength of these tools lies in transforming raw satellite data into actionable information to enable effective environmental monitoring. Hence, to demonstrate the practical utility, a long-term (2001–2021) NDVI-based land categorization of India using the cloud-based EO platform of Google Earth Engine (GEE) was carried out. The study analyzed mean NDVI values from 2001 to 2021 to assess vegetation trends and land category dynamics over two decades.

**Table 2.4** Commonly used web-based EO data integration platforms

| Sl. No | Platform | Key features | Data sources | User interface |
|---|---|---|---|---|
| 1 | Google Earth Engine | Petabyte-scale storage; supports JavaScript and Python APIs; global EO archives; cloud processing (Mutanga & Kumar, 2019) | Landsat, Sentinel, MODIS, CHIRPS, etc. | Web-based IDE; Python/JS API |
| 2 | ESA Copernicus DIAS | Access to Copernicus datasets; multi-user cloud access; processing-as-a-service via DIAS partners (Batina, 2020) | Sentinel-1, Sentinel-2, Sentinel-5P, etc. | Web portal with DIAS partner services |
| 3 | Sentinel Hub | Fast WMS/WMTS streaming; EO Browser; API-based access to Sentinel imagery (Milcinski & Kolaric, 2023) | Sentinel missions | EO Browser; API interface |
| 4 | NASA Earth data & AppEEARS | Search, subset, download EO datasets; AppEEARS allows spatial/temporal extraction and ready-to-use data (Shen et al., 2013) | MODIS, VIIRS, SMAP, Landsat, etc. | Web portals; AppEEARS GUI |
| 5 | AWS Open Data Program | Open access to EO datasets; integration with AWS tools for large-scale parallel processing (Rizvi et al., 2018) | Landsat, Sentinel, NEXRAD, etc. | AWS CLI & Web console |
| 6 | Microsoft Planetary Computer | STAC-compliant data catalog; JupyterHub; Azure ML integration for EO and climate applications (Lukacz et al., 2022) | Harmonized Landsat-Sentinel, NAIP, MS imagery | Jupyter Notebooks & REST APIs |

#### 2.3.2.1 Long-Term Mean NDVI-Based Land Categorization of India

A long-term vegetation assessment was conducted for the entire stretch of India in order to exemplify the application of web-based EO platforms for large-scale environmental monitoring using Google Earth Engine (GEE). This case study employed annually averaged NDVI values over a 21-year period (2001–2021) to classify the landscape into distinct land categories based on vegetation density and coverage. Preprocessed datasets from the USGS Landsat 5 Level 2, Collection 2, Tier 1, and USGS Landsat 8 Collection 2, Tier 1 archives were accessed using JavaScript APIs within the GEE environment. Following data acquisition, NDVI values were computed using the spectral bands designated for red and near-infrared reflectance (as given in Table 2.3), and a spatial resampling was applied to achieve a uniform resolution of 2,500 m. Subsequently, NDVI-based classification was performed using predefined threshold values to delineate land categories and is shown in Fig. 2.4.

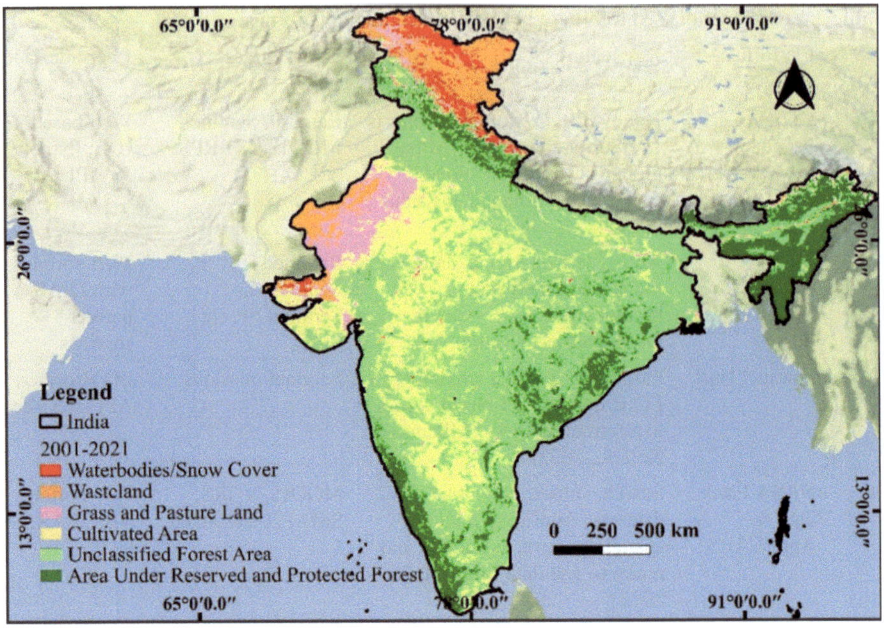


**Fig. 2.4** Long-term NDVI-based land categorization of India (2001–2021)

Results of the analysis further revealed that regions with the highest NDVI values corresponding to dense vegetation and forest cover were predominantly located in the northeastern states, the Western Ghats, and the Himalayan foothills of Uttarakhand and Himachal Pradesh. In contrast, the central and northwestern parts of India showed extensive coverage under cultivated areas. A noticeable reduction in wasteland and grassland/pasture categories was observed when compared to annual land categorization maps. This is likely due to the smoothing effect of long-term averaging which reduces the impact of inter-annual variability and seasonal fluctuations (Gogineni & Chintalacheruvu, 2024; Goyal & Rautela, 2024). Thus, this assessment highlights that relying solely on single-year NDVI data may lead to misleading interpretations due to temporal anomalies or transient land use practices. Instead, long-term NDVI trends provide a more robust and reliable basis for identifying persistent land cover patterns. The areal distribution of each land category and their state-wise statistics for the years 2001, 2021, and the aggregated period 2001–2021 are further shown in Fig. 2.5 to emphasize both spatial and temporal shifts in land categories.

Thus, the contrast between annual and long-term NDVI threshold-based land categorization of India is distinctly shown in Fig. 2.5a. Desktop-based analysis using annual EO datasets for the years 2001 reveals that the western states of Gujarat and Rajasthan exhibited significantly low mean NDVI values ranging from 0.047 to 0.386. This indicates sparse vegetation and dominant arid landscapes. Similarly, the central states of Madhya Pradesh and Maharashtra exhibited low NDVI values

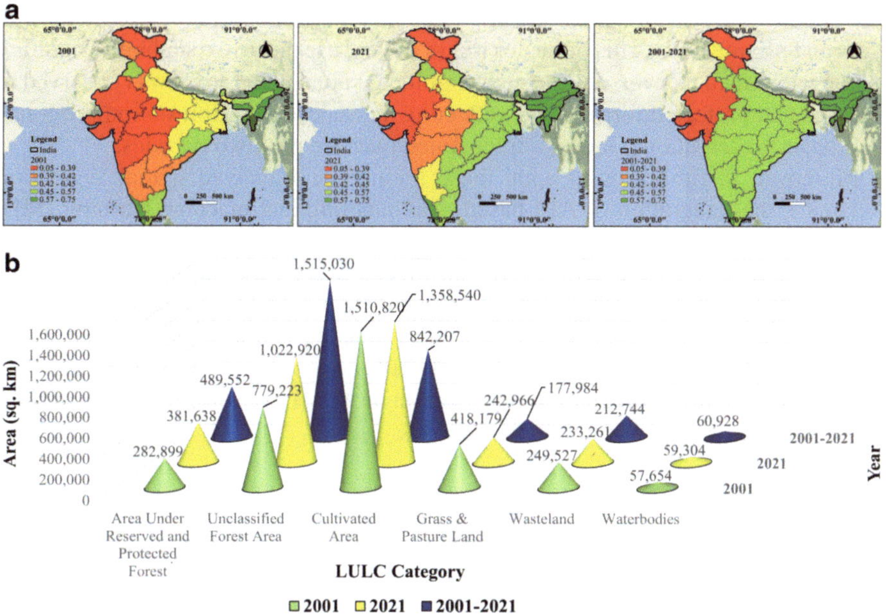


**Fig. 2.5** **a** Year-wise NDVI-based categorization of Indian states; **b** Areal extent of each land category

in 2001; however, by 2021, these states demonstrated improved vegetation conditions with mean NDVI values increasing to the range of 0.386–0.421. Likewise, the southern states of Telangana, Karnataka, and Andhra Pradesh, which recorded moderate vegetation cover in 2001 (NDVI range: 0.386–0.421), experienced an upward shift in vegetation by 2021. Specifically, Telangana and Andhra Pradesh reached NDVI values of 0.451–0.574 while Karnataka moved to the 0.421–0.451 range. The eastern states of Chhattisgarh, West Bengal, Jharkhand, and Odisha exhibited similar increase in mean NDVI values. Notably, Odisha maintained a relatively stable NDVI range of 0.421–0.451 across both years. Similar observation could be made for the Northeastern states characterized by dense forest cover and consistently high NDVI values in both years. Overall, the annual mean NDVI-based categorization reflects a positive trend in vegetation improvement across most Indian states between 2001 and 2021.

However, a contrasting pattern emerges from the long-term mean NDVI analysis (2001–2021). Except for regions with inherent arid or snow-dominated conditions such as Gujarat, Rajasthan, Delhi, Himachal Pradesh, and Ladakh, the remaining states exhibited higher long-term average NDVI values. This suggests broader vegetative stability and resilience. Hence, this observation underscores the limitations of annual or short-term assessments which are incapable of capturing enduring ecological changes or agricultural cycles and can therefore result in misleading interpretations. Further, the areal analysis of land categories as shown in Fig. 2.5b also supports these findings. Between 2001 and 2021, the maximum areal loss (42%) occurred in grass and pasturelands. This indicates a shift in land use due to expanding agriculture and development. Conversely, the most significant areal gain (35%) was observed in reserved and protected forest areas, reflecting the success of afforestation programs and forest conservation policies implemented over the past two decades.

Moreover, to further build upon these findings and analyze spatial vegetation dynamics, the integration of API-based data handling and machine learning (ML) techniques becomes indispensable. While traditional NDVI classification methods provide valuable baseline assessments, advanced ML-based clustering approaches allow for deeper pattern recognition, regional differentiation, and the identification of vegetation behavior clusters across all the states. Thus, leveraging cloud-based platforms and programmatic access through APIs not only automates large-scale EO data workflows but also enables the application of AI/ML models for more insightful and data-driven environmental interpretations.

### 2.3.3 API-Driven and AI/ML-Based EO Dataset Integration Frameworks

Building on the capabilities of advanced API-enabled and AI/ML-integrated EO data frameworks, this section presents a practical case study on unsupervised machine learning-based clustering of Indian states using long-term mean NDVI values (2001–2021). This case study highlights the real-world application of ML techniques to derive meaningful insights from EO datasets to enable data-driven environmental analysis. The state-wise long-term mean NDVI values were compiled and used as input for clustering. Prior to applying the clustering algorithm, a silhouette analysis (Gururani et al., 2023; Wang et al., 2017) was performed to determine the optimal number of clusters for the unlabeled data. This analysis was conducted in Python using the scikit-learn library's silhouette_score function (Roy et al., 2025; Shahapure & Nicholas, 2020). The results of the analysis revealed that four clusters provided the most distinct and internally coherent grouping. The silhouette scores across varying cluster numbers are shown in Fig. 2.6a. The highest score was observed at four clusters which is indicative of the most meaningful separation between mean NDVI values.

Thus, Table 2.5 presents the core capabilities and typical use cases of widely used API-driven and AI/ML-based EO integration frameworks. However, their true potential becomes evident through practical implementation. Further, EO data workflows in integration with machine learning-based analysis enable researchers and decision-makers to derive actionable insights tailored to different geographic regions and thematic applications.

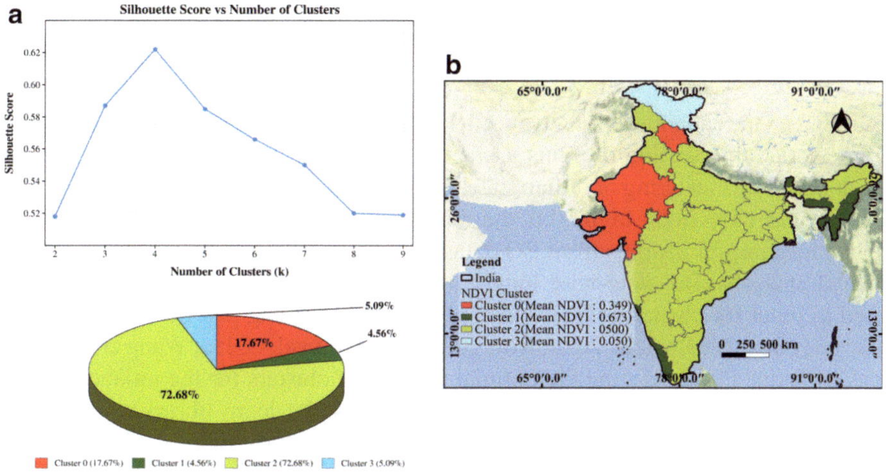


**Fig. 2.6** **a** Silhouette score test results; **b** Long-term mean NDVI-based cluster zones of India

**Table 2.5** Commonly utilized API-driven And AI/ML-based EO integration frameworks

| Sl. No | Platform/ framework | Core capabilities | Common use cases | References |
|---|---|---|---|---|
| 1 | Google Earth Engine JS and Python (API) | Planet-scale geospatial processing; supports Python/JS; automated classification & change detection | Deforestation monitoring, drought modeling, urban sprawl analysis | Gorelick et al. (2017) |
| 2 | Sentinel Hub API | Customizable satellite imagery access via WMS/WMTS; cloud masking; statistical summaries | NDVI trends, land classification, seasonal EO visualization | Loukili et al. (2022) |
| 3 | AWS Lambda + Earth on AWS | Serverless cloud computing for scalable geospatial pipelines; integrates with S3-hosted EO data | Real-time flood mapping, image tiling, batch processing | Schnase et al. (2017) |
| 4 | Microsoft Planetary Computer | STAC-compatible dataset access; Azure ML integration; spatiotemporal data fusion | Land change analysis, biodiversity studies, climate model input | Microsoft (2022) |
| 5 | Google Colab (AI/ML) | Interactive ML model training for EO data; image segmentation and regression | Crop classification, forest segmentation, yield prediction | Chen et al. (2023) |
| 6 | Azure AI for Earth | Cloud-based AI model deployment for EO applications; automated geospatial workflows | Disaster response modeling, conservation monitoring | Sánchez (2024) |
| 7 | Jupyter + scikit-learn/ TensorFlow | Customizable AI/ML pipelines for EO data using Python libraries; reproducible modeling | Air quality modeling, land use forecasting, temporal trend detection | Le Piane et al. (2024) |

#### 2.3.3.1 ML-Based Clustering of Indian States Using Long-Term Mean NDVI Values

Building on the capabilities of advanced API-enabled and AI/ML-integrated EO data frameworks, this section presents a practical case study on unsupervised machine learning-based clustering of Indian states using long-term mean NDVI values (2001–2021). This case study highlights the real-world application of ML techniques to derive meaningful insights from EO datasets, thereby enabling data-driven environmental analysis. The state-wise long-term mean NDVI values were compiled and used as input for clustering. Prior to applying the clustering algorithm, a silhouette analysis (Goyal & Rautela, 2024; Roy, 2024; Roy & Goyal, 2025a; Wang et al., 2017) was performed to determine the optimal number of clusters for the unlabeled data. This analysis was conducted in Python using the scikit-learn library's silhouette_score function, revealed that four clusters provided the most distinct and internally coherent grouping. The silhouette scores across varying cluster numbers are shown in Fig. 2.6a, where the highest score is observed at four clusters, indicating the most meaningful separation between states.

The silhouette method ensures that the defined clusters are both internally consistent (high intra-cluster similarity) and externally distinct (low inter-cluster similarity) (Roy & Chintalacheruvu, 2024a, 2024b, 2024c). Importantly, this type of unsupervised ML approach allows for EO-based pattern detection without requiring prior labeling or training data, making it suitable for large-scale, exploratory environmental assessments. Following the silhouette analysis, the K-means clustering algorithm was applied to classify the Indian states into four distinct clusters. The resulting cluster-wise mean NDVI values were also computed and are shown in Fig. 2.6b (Ahmad & Dey, 2007; Roy & Chintalacheruvu, 2024a, 2024b, 2024c; Singh & Kumar, 2024). The analysis identified the following clusters:

**Cluster 3 (Mean NDVI = 0.050)**: Occupying only 5.09% of India's total land area, this cluster comprises snow-dominated regions with minimal or no vegetative cover, primarily in Ladakh and high-altitude zones.

**Cluster 1 (Mean NDVI = 0.673)**: Encompassing 4.56% of India's land area, this cluster includes the dense forested regions of the Northeastern states, the Western Ghats, and parts of Kerala and Goa. These ecologically sensitive zones play a critical role in global carbon exchange and ecosystem balance.

**Cluster 0 (Mean NDVI = 0.349)**: Covering 17.67% of the land area, this cluster includes the arid regions of Gujarat and Rajasthan, the urbanized capital of Delhi, and the snow-prone state of Himachal Pradesh.

**Cluster 2 (Mean NDVI = 0.500)**: Representing the majority (72.68%) of India's landmass, this cluster covers states with moderate vegetation, such as Madhya Pradesh, Maharashtra, Chhattisgarh, and parts of the Deccan Plateau (Rautela et al., 2024a, 2024b; Rakkasagi et al., 2025, Singh et al., 2023).

The results demonstrate that K-means clustering, when applied to long-term EO data, can effectively group regions based on vegetation intensity and land cover stability. These spatially informed clusters provide a scientific foundation for environmental zoning, resource planning, and ecological policy development. The workflow illustrates how API-driven and AI/ML-integrated EO analysis enhances the generation of actionable insights and supports evidence-based decision-making. The next phase in this evolution involves expanding the application of machine learning across diverse EO domains to enable more precise, scalable, and automated modeling. This ongoing integration of intelligent algorithms marks a significant shift toward adaptive, data-driven approaches in integrated EO and ML based environmental monitoring.

## 2.4 Conclusions

Effective utilization of Earth Observation (EO) data depends on structured approaches to acquisition, preprocessing, integration, and analysis. As EO datasets continue to grow in volume, frequency, and diversity, robust data management strategies and scalable analytical tools are essential to ensure accurate and reproducible environmental insights. Standardized data processing levels, ranging from Level 0 to Level 4, enable consistent formatting, usability, and long-term accessibility of EO products. Analysis of long-term NDVI values across India demonstrated the advantages of temporal averaging in revealing persistent vegetation patterns, which are often masked in single-year assessments. The long-term mean NDVI-based classification highlighted that the northeastern states, the Western Ghats, and Himalayan foothills maintain the highest vegetative cover, while arid regions and urban centers continue to show low NDVI values.

Significant observations included a 42% decline in grass and pasture lands and a 35% increase in reserved and protected forests between 2001 and 2021, suggesting notable land use transitions and the possible impact of conservation efforts. To extract more actionable patterns from this EO-derived information, unsupervised machine learning techniques were employed. K-means clustering effectively grouped Indian states into four ecological zones based on their mean NDVI values, offering a data-driven foundation for environmental zoning, vegetation monitoring, and resource planning. Web-based EO platforms such as Google Earth Engine, Sentinel Hub, and AWS Earth on Cloud, combined with API integrations and AI/ML algorithms, enable the automation of large-scale geospatial analyses. These platforms facilitate near-real-time processing, cloud storage, and seamless access to multi-source datasets, transforming the way environmental change is monitored, modeled, and managed. As environmental systems grow increasingly complex, the integration of EO data with intelligent analytical frameworks offers a powerful toolset for addressing challenges such as climate change, land degradation, and ecosystem management. Continued innovation in machine learning, cloud computing, and EO data integration is driving the shift toward more adaptive, automated, and evidence-based environmental monitoring solutions.

## References


Ahmad, A., & Dey, L. (2007). A k-mean clustering algorithm for mixed numeric and categorical data. *Data & Knowledge Engineering, 63*(2), 503–527. https://doi.org/10.1016/j.datak.2007.02.003

Albani, S., & Giaretta, D. (2009). Long-term preservation of earth observation data and knowledge in ESA through CASPAR. *International Journal of Digital Curation, 4*(3), 4–16. https://doi.org/10.2218/ijdc.v4i3.123

Allen, D. W. (2011). *Getting to know ArcGIS ModelBuilder*. Esri Press.

Anand, V., & Oinam, B. (2022). Modeling the potential impact of land use/land cover change on the hydrology of Himalayan River Basin: A case study of Manipur River, India. In *Handbook of Himalayan ecosystems and sustainability* (Vol. 2, pp. 189–204). CRC Press.

Bannari, A., Morin, D., Bénié, G. B., & Bonn, F. J. (1995). A theoretical review of different mathematical models of geometric corrections applied to remote sensing images. *Remote Sensing Reviews, 13*(1–2), 27–47. https://doi.org/10.1080/02757259509532283

Batina, A. (2020). Copernicus services and DIAS—Evaluation of efficiency and applicability. *Geodetski List, 74*(3), 315–336.

Chen, H., Yang, L., & Wu, Q. (2023). Enhancing land cover mapping and monitoring: An interactive and explainable machine learning approach using Google Earth Engine. *Remote Sensing, 15*(18), 4585. https://doi.org/10.3390/rs15184585

Cheng, Y., Zhou, K., Wang, J., & Yan, J. (2020). Big earth observation data integration in remote sensing based on a distributed spatial framework. *Remote Sensing, 12*(6), 972. https://doi.org/10.3390/rs12060972

Congedo, L. (2016). *Semi-automatic classification plugin documentation* (Release 4.0.1). https://semiautomaticclassificationmanual.readthedocs.io

Doxani, G., Vermote, E. F., Roger, J. C., Skakun, S., Gascon, F., Collison, A., & Yin, F. (2023). Atmospheric correction Inter-comparison eXercise, ACIX-II Land: An assessment of atmospheric correction processors for Landsat 8 and Sentinel-2 over land. *Remote Sensing of Environment, 285*, Article 113412. https://doi.org/10.1016/j.rse.2022.113412

Duane, W., Livingstone, D., & Kidd, D. (2000). Integrating environmental models with GIS: An object-oriented approach utilising a hierarchical data format (HDF) data repository. *Transactions in GIS, 4*(3), 263–280. https://doi.org/10.1111/1467-9671.00052

Eidenshink, J. C., & Faundeen, J. L. (1994). The 1 km AVHRR global land data set: First stages in implementation. *International Journal of Remote Sensing, 15*(17), 3443–3462. https://doi.org/10.1080/01431169408954352

Folk, M., McGrath, R. E., & Yeager, N. (1999, June). HDF: An update and future directions. In *IEEE 1999 International Geoscience and Remote Sensing Symposium. IGARSS'99* (Vol. 1, pp. 273–275). IEEE. https://doi.org/10.1109/IGARSS.1999.773481

Gogineni, A., & Chintalacheruvu, M. R. (2024). Assessing temporal snow cover variation in the Sutlej river basin using Google Earth Engine and machine learning models. *Earth Science Informatics, 17*(1), 455–473.

Gomes, V. C., Queiroz, G. R., & Ferreira, K. R. (2020). An overview of platforms for big earth observation data management and analysis. *Remote Sensing, 12*(8), 1253. https://doi.org/10.3390/rs12081253

Gong, P., Wang, J., Yu, L., Zhao, Y., et al. (2020). Global annual maps of fine-resolution land cover for 2016 and 2017. *Nature Sustainability, 3*, 505–513. https://doi.org/10.1038/s41893-020-0523-3

Gorelick, N., Hancher, M., Dixon, M., Ilyushchenko, S., Thau, D., & Moore, R. (2017). Google Earth Engine: Planetary-scale geospatial analysis for everyone. *Remote Sensing of Environment, 202*, 18–27. https://doi.org/10.1016/j.rse.2017.06.031

Gowan, T. A., Horel, J. D., Jacques, A. A., & Kovac, A. (2022). Using cloud computing to analyze model output archived in Zarr format. *Journal of Atmospheric and Oceanic Technology, 39*(4), 449–462. https://doi.org/10.1175/JTECH-D-21-0041.1

Goyal, M. K., & Rautela, K. S. (2024). Aerosol atmospheric rivers: Detection and spatio-temporal patterns. In *Aerosol atmospheric rivers: Availability, spatiotemporal characterisation, predictability, and impacts* (pp. 19–41). Cham: Springer Nature Switzerland.

Goyal, M. K., Rakkasagi, S., Surampalli, R. Y., Zhang, T. C., Erumalla, S., Gupta, A., & U-tapao, C. (2024). Enhancing sustainable development through Spatiotemporal analysis of Ramsar wetland sites in South Asia. *Technology in Society, 79*, Article 102723.

Gururani, D. M., Singh, S., Joshi, H., Kumar, Y., Kumar, A., Bohra, M. S., & Mehta, P. (2023). Assessment of groundwater prospects zones using RS, GIS, and MIF methods. In *Surface and groundwater resources development and management in semi-arid region: Strategies and solutions for sustainable water management* (pp. 317–335). Springer International Publishing.

Hansen, M. C., Potapov, P. V., Moore, R., et al. (2013). High-resolution global maps of 21st-century forest cover change. *Science, 342*(6160), 850–853. https://doi.org/10.1126/science.1244693

Hinge, G., Surampalli, R. Y., & Goyal, M. K. (2018). Regional carbon fluxes from land-use conversion and land-use management in Northeast India. *Journal of Hazardous, Toxic, and Radioactive Waste, 22*(4), 04018016.

Hostetler, C. A., Behrenfeld, M. J., Hu, Y., Hair, J. W., & Schulien, J. A. (2018). Spaceborne lidar in the study of marine systems. *Annual Review of Marine Science, 10*(1), 121–147.

Hu, K., Yang, X., Zhong, J., Fei, F., & Qi, J. (2017). Spatially explicit mapping of heat health risk utilizing environmental and socioeconomic data. *Environmental Science & Technology, 51*(3), 1498–1507. https://doi.org/10.1021/acs.est.6b03342

Jain, V., Kumar, S., & Goyal, M. K. (2025). Relationship between daily precipitation extremes and temperature in changing climate across smart cities of Central India. *Journal of Environmental Management, 380*, Article 125036.

Jain, V., Rautela, K. S., & Goyal, M. K. (2023). Ecological restoration: An overview of science and policy regime. *Ecosystem Restoration: Towards Sustainability and Resilient Development*, 1–27.

Kennedy, C. M., Oakleaf, J. R., Theobald, D. M., Baruch-Mordo, S., & Kiesecker, J. (2019). Managing the middle: A shift in conservation priorities based on the global human modification gradient. *Global Change Biology, 25*(3), 811–826. https://doi.org/10.1111/gcb.14549

Khan, J., & Ramesh, R. (2024). Performance evaluation of geospatial images based on Zarr and TIFF. *arXiv Preprint*. arXiv:2411.11291. https://arxiv.org/abs/2411.11291

Khan, M., Sharma, A., & Goyal, M. K. (2019). Assessment of future water provisioning and sediment load under climate and LULC change scenarios in a Peninsular River Basin, India. *Hydrological Sciences Journal, 64*(4), 405–419.

Le Piane, F., Vozza, M., Baldoni, M., & Mercuri, F. (2024). Integrating high-performance computing, machine learning, data management workflows, and infrastructures for multiscale simulations and nanomaterials technologies. *Beilstein Journal of Nanotechnology, 15*(1), 1498–1521.

Li, L., Zhang, G., Jiang, Y., & Shen, X. (2021). An improved on-orbit relative radiometric calibration method for agile high-resolution optical remote-sensing satellites with sensor geometric distortion. *IEEE Transactions on Geoscience and Remote Sensing, 60*, 1–15. https://doi.org/10.1109/TGRS.2021.3112192

Loukili, Y., Lakhrissi, Y., & Ali, S. E. B. (2022). Geospatial big data platforms: A comprehensive review. *KN-Journal of Cartography and Geographic Information, 72*(4), 293–308.

LTDP Working Group. (2013). *Long term preservation of earth observation space data* (Vol. 2). European Space Agency.

Lukacz, P. M. (2022, February). Data capitalism, Microsoft's planetary computer, and the biodiversity informatics community. In *International Conference on Information* (pp. 355–369). Cham: Springer.

Lukacz, P. M. (2024). Imaginaries of democratization and the value of open environmental data: Analysis of Microsoft's Planetary Computer. *Big Data & Society, 11*(2), 20539517241242450. https://doi.org/10.1177/20539517241242448

Mahiny, A. S., & Turner, B. J. (2007). A comparison of four common atmospheric correction methods. *Photogrammetric Engineering & Remote Sensing, 73*(4), 361–368. https://doi.org/10.14358/PERS.73.4.361

Microsoft. (2022). Planetary computer documentation. *Microsoft AI for Earth.* Retrieved from https://planetarycomputer.microsoft.com/docs/

Milcinski, G., & Kolaric, P. (2023, May). Sentinel Hub–federated on-demand ARD generation. In *EGU General Assembly Conference Abstracts* (pp. EGU-4160).

Molthan, A. L., Case, J. L., Venner, J., Schroeder, R., Checchi, M. R., Zavodsky, B. T., & O'Brien, R. G. (2015). Clouds in the cloud: Weather forecasts and applications within cloud computing environments. *Bulletin of the American Meteorological Society, 96*(8), 1369–1379.

Mutanga, O., & Kumar, L. (2019). Google Earth Engine applications. *Remote Sensing, 11*(5), 591.

NASA Earthdata. (2025). Data processing levels. https://www.earthdata.nasa.gov/learn/earth-observation-data-basics/data-processing-levels

Okolie, C. J., & Smit, J. L. (2022). A systematic review and meta-analysis of digital elevation model (DEM) fusion: Pre-processing, methods and applications. *ISPRS Journal of Photogrammetry and Remote Sensing, 188*, 1–29. https://doi.org/10.1016/j.isprsjprs.2022.03.007

Paolini, L., Grings, F., Sobrino, J. A., Jiménez-Muñoz, J. C., & Karszenbaum, H. (2006). Radiometric correction effects in Landsat multi-date/multi-sensor change detection studies. *International Journal of Remote Sensing, 27*(4), 685–704. https://doi.org/10.1080/01431160500162469

Pérez-Peña, J. V., Azañón, J. M., & Azor, A. (2009). CalHypso: An ArcGIS extension to calculate hypsometric curves and their statistical moments. Applications to drainage basin analysis in SE Spain. *Computers & Geosciences, 35*(6), 1214–1223.

Rakkasagi, S., Edris, S. G., & Goyal, M. K. (2025). A socio-technical framework for assessing flash drought risks to key biodiversity areas: A scientists' warning. *Technology in Society, 82*, Article 102893.

Rautela, K. S., Goyal, M. K., & Surampalli, R. Y. (2025). AI and machine learning for optimizing waste management and reducing air pollution. *Journal of Hazardous, Toxic, and Radioactive Waste, 29*(3), 04025014.

Rautela, K. S., Singh, S., & Goyal, M. K. (2024a). Characterizing the spatio-temporal distribution, detection, and prediction of aerosol atmospheric rivers on a global scale. *Journal of Environmental Management, 351*, Article 119675.

Rautela, K. S., Singh, S., & Goyal, M. K. (2024b). Resilience to air pollution: A novel approach for detecting and predicting aerosol atmospheric rivers within earth system boundaries. *Earth Systems and Environment*, 1–23.

Rizvi, S. R., Killough, B., Cherry, A., & Gowda, S. (2018, July). Lessons learned and cost analysis of hosting a full stack Open Data Cube (ODC) application on the Amazon Web Services (AWS). In *IGARSS 2018—IEEE International Geoscience and Remote Sensing Symposium* (pp. 8643–8646). IEEE.

Roy, S. (2024). Decoding the declining mangrove ecosystems: An analysis of carbon dynamics, salinity changes, and deforestation using remote sensing and machine learning. In *AGU24.*

Roy, S., & Chintalacheruvu, M. R. (2023, February). LULC dynamics study and modeling of urban land expansion using CA-ANN. In *International Conference on Science, Technology and Engineering* (pp. 79–90). Singapore: Springer Nature Singapore.

Roy, S., & Chintalacheruvu, M. R. (2024a). Enhanced morphometric analysis for soil erosion susceptibility mapping in the Godavari River basin, India: Leveraging Google Earth Engine and principal component analysis. *ISH Journal of Hydraulic Engineering, 30*(2), 228–244.

Roy, S., & Chintalacheruvu, M. R. (2024b). Google Earth Engine-based morphometric parameter evaluation and comparative analysis of soil erosion susceptibility using statistical and machine learning algorithms in large river basins. *Earth Science Informatics, 17*(1), 75–97. https://doi.org/10.1007/s12145-023-00946-7

Roy, S., & Chintalacheruvu, M. R. (2024c). Delineating hydro-geologically constrained groundwater zones in the Himalayan River basins of India through an innovative ensemble of

hypsometric analysis and machine learning algorithms. *Earth Science Informatics, 17*(1), 501–526.
Roy, S., & Goyal, M. (2025). Climate variability and extremes in Indian River Basins: Trends, regional disparities, and urban risk assessment (No. EGU25-1114). In *Copernicus Meetings.*
Roy, S., Rao, C. M., & Abioui, M. (2024). Evaluation of non-stationary spatial relationship between meteorological-environmental parameters and PM2.5. *Advances in Space Research, 73*(8), 4106–4124.
Roy, S., Uddin, M. G., Abdelrahman, K., Fnais, M. S., & Abioui, M. (2025). Assessing the impact of digital elevation model resolution on hypsometric analysis in large river Basins (India): A non-parametric statistical approach. *Earth Science Informatics, 18*(1), 1–21.
Sánchez, A. G. (2024). *Azure OpenAI service for cloud native applications.* O'Reilly Media, Inc.
Schnase, J. L., Duffy, D. Q., Tamkin, G. S., Nadeau, D., Thompson, J. H., Grieg, C. M., & Webster, W. P. (2017). MERRA analytic services: Meeting the big data challenges of climate science through cloud-enabled climate analytics-as-a-service. *Computers, Environment and Urban Systems, 61*, 198–211.
Shahapure, K. R., & Nicholas, C. (2020, October). Cluster quality analysis using silhouette score. In *2020 IEEE 7th International Conference on Data Science and Advanced Analytics (DSAA)* (pp. 747–748). IEEE.
Shangguan, W., Dai, Y., Duan, Q., Liu, B., & Yuan, H. (2014). A global soil data set for earth system modeling. *Journal of Advances in Modeling Earth Systems, 6*(1), 249–263. https://doi.org/10.1002/2013MS000293
Shen, S., Leptoukh, G., & Fang, H. (2013). NASA satellite and model land data services: Data access tutorial. In *Land surface observation, modeling and data assimilation* (pp. 67–89).
Singh, S., & Kumar, A. (2024). Understanding the intricacies of rainfall dynamics using entropy measures. *Journal of Water and Climate Change, 15*(9), 4817–4839.
Singh, S., Kumar, D., Kumar, A., & Kuriqi, A. (2023). Entropy-based assessment of climate dynamics with varying elevations for hilly areas of Uttarakhand, India. *Sustainable Water Resources Management, 9*(4), 130.
Sonka, M., Hlavac, V., & Boyle, R. (2013). *Image processing, analysis and machine vision* (4th ed.). Springer.
Sudmanns, M., Tiede, D., Lang, S., Bergstedt, H., Trost, G., Augustin, H., & Blaschke, T. (2020). Big Earth data: Disruptive changes in earth observation data management and analysis? *International Journal of Digital Earth, 13*(7), 832–850. https://doi.org/10.1080/17538947.2019.1585976
Tambe, S., Rawat, G. S., Krishen, P., Ranjitsinh, M. K., Ghosh, N., Rawat, A. S., & Takpa, J. (2022). Compensatory afforestation policy in India: An analysis using an ecorestoration lens. *International Forestry Review, 24*(4), 607–618.
Thaler, S. (2016). Storing NetCDF data in a relational database. *Zenodo.* https://doi.org/10.5281/zenodo.61147
Troumpoukis, A., Klampanos, I., Pantazi, D. A., Albughdadi, M., Baousis, V., Barrilero, O., & Karkaletsis, V. (2024). European AI and EO convergence via a novel community-driven framework for data-intensive innovation. *Future Generation Computer Systems, 160*, 505–521.
Vitalis, S., Arroyo Ohori, K., & Stoter, J. (2020). CityJSON in QGIS: Development of an open-source plugin. *Transactions in GIS, 24*(5), 1147–1164.
Wang, F., Franco-Penya, H. H., Kelleher, J. D., Pugh, J., & Ross, R. (2017). An analysis of the application of simplified silhouette to the evaluation of k-means clustering validity. In *Machine Learning and Data Mining in Pattern Recognition: 13th International Conference, MLDM 2017* (pp. 291–305). Springer International Publishing.
Waring, R. H., Way, J., Hunt, E. R., Morrissey, L., Ranson, K. J., Weishampel, J. F., & Franklin, S. E. (1995). Imaging radar for ecosystem studies. *BioScience, 45*(10), 715–723.
Wilkinson, R., Mleczko, M. M., Brewin, R. J. W., Gaston, K. J., Mueller, M., Shutler, J. D., & Anderson, K. (2024). Environmental impacts of earth observation data in the constellation and cloud computing era. *Science of the Total Environment, 909*, Article 168584. https://doi.org/10.1016/j.scitotenv.2023.168584

Wulder, M. A., Loveland, T. R., Roy, D. P., et al. (2018). Current status of Landsat program, science, and applications. *Remote Sensing of Environment, 225*, 127–147. https://doi.org/10.1016/j.rse.2019.02.015

Xue, L. (2010, October). Application of IDL and ENVI redevelopment in hyperspectral image preprocessing. In *International Conference on Computer and Computing Technologies in Agriculture* (pp. 403–409). Berlin, Heidelberg: Springer.

Zhu, Zhe, & Woodcock, Curtis E.. (2012). Object-based cloud and cloud shadow detection in Landsat imagery. *Remote Sensing Of Environment, 118*, 83–94. https://doi.org/10.1016/j.rse.2011.10.028

Zhuk, E. (2024, September). Web GIS tools to combine station observation and netCDF format data. In *Tenth International Conference on Remote Sensing and Geoinformation of the Environment (RSCy2024)* (Vol. 13212, pp. 495–499). SPIE. https://doi.org/10.1117/12.3186710

# Chapter 3
# Machine Learning for Earth Observation

## 3.1 Introduction to Machine Learning in Earth Observation

ML is a branch of artificial intelligence that learns and improves its performance by implementing computational methodologies. Here, it is worth mentioning that the term "learn" in computer terminology refers to the storage of these "learnings" or "experiences" in the form of data. Thus, the final and crucial utilization of these learnings is applied in developing models capable of making accurate predictions by simulating real scenarios (El Naqa & Murphy, 2015; Zhou, 2021). Hence, ML algorithms find their utilization in almost all aspects of EO dataset analysis for environmental monitoring. Algorithms like unsupervised learning (e.g., K-means clustering, hierarchical clustering etc.) and supervised classification (e.g., random forest, decision tree, support vector machine, etc.) are implemented for LULC classification (Anand & Oinam, 2020; Hasmadi et al., 2009; Roy & Chintalacheruvu, 2023a, 2023b), hydrological modeling (Iqbal et al., 2022; Yang et al., 2020), water quality analysis (Leggesse et al., 2023; Mukonza & Chiang, 2023; Peterson et al., 2020), air quality analysis (Lary et al., 2015), and overall environmental monitoring (Hino et al., 2018; Lary et al., 2018; Persello et al., 2022).

## 3.2 Machine Learning Applications in Environmental Monitoring

ML algorithms now regularly find their utilization in environmental monitoring and in the analysis of EO datasets obtained from satellites and sensors and saved in different formats like vector, raster, images, or records. ML algorithms enable the creation of continuous and homogeneous datasets using regression algorithms

 
S. Roy and M. K. Goyal, *Environmental Monitoring with Integrated Earth Observation Data and Machine Learning*, Studies in Computational Intelligence 1278,
https://doi.org/10.1007/978-3-032-19935-5_3

(Abdul-Wahab et al., 2005; Coskun et al., 2008) further enabling the prediction of future scenarios essential for LULC studies (Anand & Oinam, 2020) and climate analysis (Dey et al., 2022; Roy & Goyal, 2025). Similarly, unsupervised ML algorithms enable the grouping of similar data points, thereby helping in the identification of the naturally existing spatial patterns. Such analysis is indispensable for assessing the impacts of water quality patterns, landform characterization, epidemic outbreak studies, etc. (Fusco et al., 2020; Rahman et al., 2023; Roy & Chintalacheruvu, 2024). A complete tabulation highlighting some of the most used ML libraries along with their application domains is given in Table 3.1.

Thus, as given in Table 3.1, there is no dearth of availability of ML algorithms and tools for environmental monitoring. However, understanding their practical utility requires examining how these algorithms are applied in real-world scenarios. Hence, to demonstrate their practical utility, ML-based methods were used to analyze long-term precipitation data of a region and implement the derived specific insights of this analysis.

### *3.2.1 ML-Based Analysis of Long-Term Precipitation in a Region*

The practical implementation of ML-based tools and algorithms for environmental monitoring purposes was demonstrated through a case study. It incorporated the analysis of long-term precipitation datasets of the state of Madhya Pradesh of central India. The input dataset was obtained from the gridded precipitation dataset of the India Meteorological Department (IMD) in NetCDF format from 1964 to 2023. The data were arranged grid-wise wherein each unique latitude and longitude pair was considered as a station. The annual precipitation observed over the entire region ranged from 633.93 to 1658.80 mm with a mean annual precipitation of 1036.33 mm. Following this, the assessment of long-term precipitation trends at each station was assessed using the two nonparametric statistical tests of modified Mann–Kendall (MMK) test for detecting significant monotonic trends in the data and the innovative trend analysis (ITA) method to capture more complex patterns in precipitation variation (Dixit & Pandey, 2024; Hamed & Rao, 1998; Şen, 2012). Stations that were statistically significant in both tests were considered for subsequent Standardized Precipitation Index (SPI) analysis. The yearly SPI values were computed to detect dry and wet anomalies across time. SPI values were further labeled as:

$$\text{Dry (SPI} \leq -1.0\text{); Normal } (-1.0 < \text{SPI} < 1.0\text{); Wet (SPI} \geq 1.0\text{)}$$

Following this, the ML algorithm of random forest classification model (Rigatti, 2017) was trained on the labeled SPI values to capture and predict temporal variability in precipitation behavior. RF is an ensemble-based decision tree classifier that constructs multiple decision trees during training and outputs the class that is

**Table 3.1** Commonly utilized ML libraries and their application domains in environmental monitoring

| ML tool/library | Programming environment | Common environmental monitoring applications | References |
|---|---|---|---|
| Scikit-learn | Python | LULC classification, air quality prediction, drought analysis | Raschka et al. (2022), Sawant et al. (2023), Ravindiran et al. (2023), Nelvidawati et al. (2024) |
| TensorFlow/Keras | Python | Deep learning for crop monitoring, vegetation health mapping, wildfire detection | Reiser et al. (2021), Kharade et al. (2023), Al-Mulla et al. (2023) |
| XGBoost | Python/R | Crop yield estimation, flood susceptibility modeling, snow depth assessments | Abedi et al. (2022), Wang et al. (2023), Wang et al. (2025) |
| LightGBM | Python/R | Soil erosion prediction, deforestation risk modeling, Forest fire analysis | Bian et al. (2023), Li et al. (2024), Nguyen et al. (2023) |
| Google Earth Engine ML API | JavaScript/ Python | Supervised/unsupervised classification, ecological assessments, air pollution forecasting, climate change studies | Tassi and Vizzari (2020), Pande (2022), Goyal et al. (2023) |
| PyTorch & Geotorch | Python | Spatiotemporal forecasting, climate trend modeling | Chowdhury and Sarwat (2024), Chowdhury and Sarwat (2022) |
| CatBoost | Python/R | Land degradation classification, forest loss prediction | Oraegbu and Jolaiya (2024), Cao et al. (2023), Ajin et al. (2024) |
| Weka | Java | Water quality classification, urban sprawl analysis | Raghavendra and Kumar (2023), Dota et al. (2015) |
| ArcGIS/QGIS ML Toolset | ArcPy/Python | Object-based image classification, supervised LULC mapping | Sagan et al. (2024), Prasad (2021), Lemenkova and Debeir (2023), Quinn et al. (2019) |

the mode of the predictions of individual trees. The classifier is known for its robustness against overfitting and its ability to handle high-dimensional and nonlinear data. In this study, the RF model was trained on the annual SPI data labeled with dry, normal, and wet classes. The model was evaluated separately for each of the four defined time periods: 1964–1993; 1994–2023; 2014–2023; 1964–2023. For each of

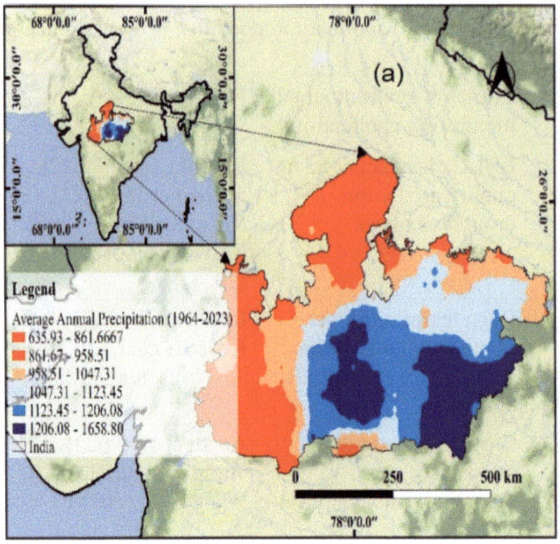


(b)

| SPI | RF | Prediction Accuracy |
|---|---|---|
| 1964-1993 | | 0.97 |
| 1967 | | |
| 1987 | | |
| 1994-2023 | | 1.00 |
| 2000 | 2000 | |
| 2013 | | |
| 2019 | | |
| 2022 | | |
| 1964-2023 | | 0.90 |
| 1965 | 1964 | |
| 1973 | | |
| 2000 | 2000 | |
| 2013 | | |
| 2019 | | |
| 2014-2023 | | 1.00 |
| 2017 | 2017 | |
| 2022 | 2022 | |

**Fig. 3.1** **a** Spatial distribution of average annual precipitation (1964–2023) in Madhya Pradesh, India; **b** Wet and dry years identified using SPI and RF-based analysis along with the prediction accuracy of the RF model

these time periods, the dataset was split into a 70:30 train–test ratio, and the model was fitted using default hyperparameters optimized through cross-validation. The performance of the classifier was measured using metrics of accuracy, precision, recall, and F1-score. The trained RF model was then used to predict wet and dry years following which they were compared against actual SPI-labeled years. This ML-driven classification allowed for better recognition of anomalous climate years across decadal and long-term time scales (Fig. 3.1).

The results of the analysis as shown in Fig. 3.1 entail the spatial distribution of precipitation across Madhya Pradesh, India. The results highlight that the central and southeastern parts of the study area received the highest average annual precipitation (1658.8 mm), while the western region remained relatively drier (633.9 mm). The RF-based classification model accurately identified anomalous precipitation years across all the study periods. Between the years 1964 and 1993, the model correctly classified 1987 and 2000 as dry years and 1967 as wet. Similarly, between 1994–2023 and 2014–2023, the dry years like 2000 and 2017, and wet years like 2013, 2019, and 2022 could be successfully identified. Over the full period (1964–2023), the RF-based model detected 1965, 1973, and 2000 as dry years, and 1964, 2013, and 2019 as wet. These results confirm the robustness of the RF model in detecting long-term precipitation anomalies. Nevertheless, this case study also emphasizes the broader potential of combining EO datasets and ML tools to support comprehensive and scalable environmental monitoring systems. Thus, the effectiveness of these ML tools significantly increases when efficiently integrated with EO platforms. This synergy between EO and ML encompasses a wide array of analytical capabilities that enable the addressing of different environmental challenges.

## 3.3 Integration of ML and EO for Environmental Monitoring

Integrated EO and ML based analysis enables the derivation of robust and data-driven insights that are scalable both spatially and temporally. This synergy also enables the detection of minute environmental anomalies along with the prediction of simulated environmental scenarios. Thus, this opens a wide range of opportunities across various domains like agricultural monitoring, infrastructure development, flood risk assessments, forest degradation, air and water quality evaluation, and even disease outbreak modeling. The scopes and practical applications of integrated EO and ML-based environmental monitoring systems are conceptually shown in Fig. 3.2. It illustrates how different EO inputs and ML algorithms converge to address diverse environmental challenges.

This integrated approach has proven effective in various domains, particularly in assessing coastal ecosystems and vegetation dynamics. Monitoring mangroves cover

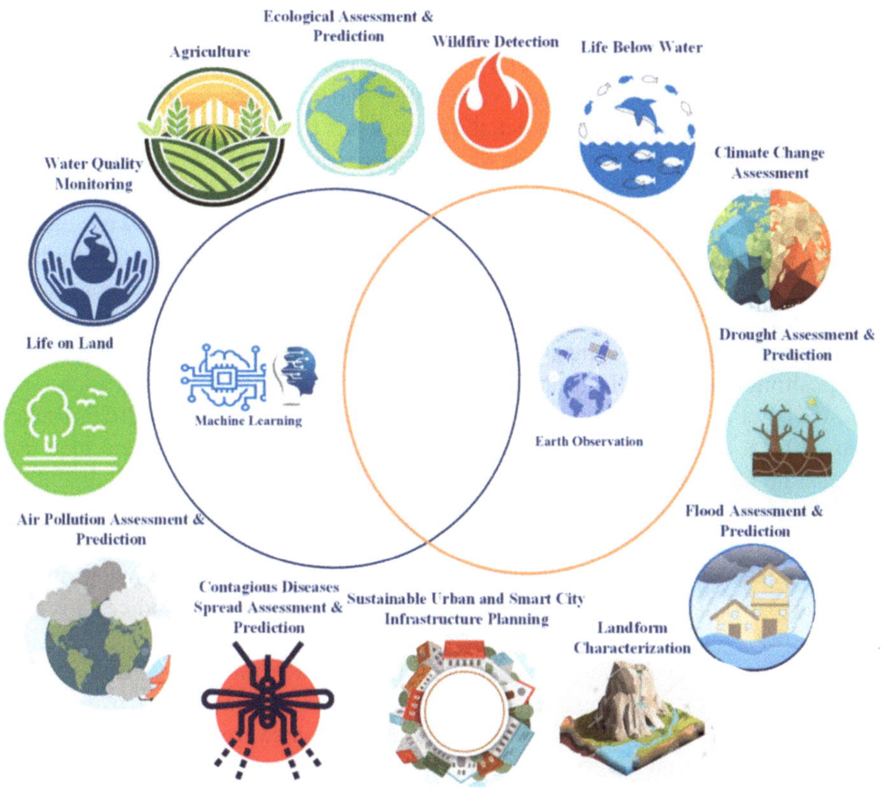


**Fig. 3.2** Conceptual representation of the synergistic applications of ML and EO for environmental monitoring

changes in sensitive coastal zones using GEE and random forest (RF) classification highlights the strength of EO-ML fusion in environmental monitoring. By combining cloud-based satellite analytics with machine learning, it enables real-time, high-resolution mapping and large-scale, long-term tracking with minimal computational effort. This demonstrates how EO and ML together can deliver precise, automated environmental insights across space and time.

### 3.3.1 EO-ML Integrated Framework for Mangrove Cover Change Detection

Understanding the dynamics of mangrove ecosystems is vital for sustainable coastal conservation and climate resilience. The Indian Sundarbans is a part of the world's largest tidal halophytic mangrove forest and a designated Ramsar site as well as a UNESCO World Heritage. This region plays a critical ecological role in coastal protection, biodiversity sustenance, and carbon sequestration (Raha et al., 2012; Roy, 2024). Spanning 6,017 $km^2$, with 40% in India and the rest in Bangladesh, this fragile ecosystem is under a lot of stress due to rising anthropogenic and climatic pressures. Thus, the computational capabilities and data storage capacities of GEE were integrated with the sophisticated data analysis techniques provided by ML to accurately map the spatial coverage of mangrove extent from 1991 to 2001.

A synergistic approach by integrating GEE with ML-based random forest (RF) classification was implemented in this study. This cloud-based framework enabled large-scale, high-resolution temporal analysis with minimal computational overhead. EO datasets from Landsat 5, Landsat 7, and Landsat 8 were utilized to generate annual means in GEE for the years of 1991, 2001, 2011, and 2021. To ensure radiometric consistency and reduce atmospheric noise, the annual means were generated by compositing data from the year immediately preceding and following each target year (Feng et al., 2022; Rakkasagi et al., 2025; Rakkasagi & Goyal, 2025; Rautela et al., 2024a, 2024b; Singh et al., 2023). These composite images were then used to generate false color composites (FCCs) in order to enhance the visual distinction between mangrove and non-mangrove areas. Subsequently, seven vegetation- and water-sensitive spectral indices were computed. These indices were used as decision tree parameters within the RF model for classifying and predicting mangrove and non-mangrove pixels. The characteristics of the selected spectral indices are given in Table 3.2.

Following the evaluation of spectral indices, the RF model was trained using these indices along with user-defined training polygons for both mangrove and non-mangrove classes for each decadal year. These polygons were split into training and testing datasets in an 80/20 ratio. The model was trained using both the spectral indices and labeled polygons. Further, it was configured with 100 decision trees to ensure strong predictive performance and minimize the risk of overfitting. This methodological setup enabled efficient and automated classification of mangrove

**Table 3.2** Characteristics of the spectral indices used for mangrove classification and spatial mapping of Sundarbans, India

| Sl. No | Index | Formula | Description | References |
|---|---|---|---|---|
| 1 | NDVI | $\frac{\text{NIR} - \text{RED}}{\text{NIR} + \text{RED}}$ | Measures green vegetation density and vigor | Meneses-Tovar (2011) |
| 2 | NDMI | $\frac{\text{NIR} - \text{SWIR1}}{\text{NIR} + \text{SWIR1}}$ | Assesses vegetation moisture content | Buzna et al. (2022) |
| 3 | MNDWI | $\frac{\text{GREEN} - \text{SWIR1}}{\text{GREEN} + \text{SWIR1}}$ | Enhance open water features while suppressing vegetation and soil noise | Yang et al. (2017) |
| 4 | SAVI | $\frac{(\text{NIR} - \text{RED})}{(\text{NIR} + \text{RED} + 0.5)*} * 1.5$ | Reduces soil brightness influence in sparse vegetation regions | Venancio et al. (2019) |
| 5 | SR | $\frac{\text{NIR}}{\text{RED}}$ | Simple ratio to distinguish vegetation from soil | Green et al. (1998) |
| 6 | MDVI | $\frac{\text{NIR} - \text{GREEN}}{\text{SWIR1} - \text{GREEN}}$ | Captures vegetation vigor and stress in mangrove ecosystems | Valderrama et al. (2018) |
| 7 | GCVI | $\frac{\text{NIR}}{\text{GREEN})} - 1$ | Indicates canopy chlorophyll content and vegetation density | Hickey and Radford (2022) |

and non-mangrove areas from 1991 to 2021. This mangrove mapping was achieved with an overall classification accuracy of 90%. Subsequently, the spatial extents of mangrove cover for each decadal year were extracted along with the mapping of changes due to deforestation and afforestation (Fig. 3.3).

The results revealed a total mangrove loss of 266.89 km$^2$ between 1991 and 2021. Out of this, the most significant deforestation occurred between 2011 and 2021 (296.37 km$^2$), followed by 58.24 km$^2$ during 1991 and 2001. In contrast, afforestation was also observed, with gains of 102.86 km$^2$, 54.47 km$^2$ and 89.95 km$^2$ during 2001–2011, 2011–2021, and 1991–2021, respectively. Thus, the results revealed 9.76% decline in mangrove cover over three decades. Compared to traditional manual or pixel-based methods, the EO-ML approach improved classification accuracy, boosted computational efficiency, and enabled seamless integration of multi-temporal and multi-source datasets (Rakkasagi et al., 2024). This integrated framework also highlights the potential of automated, data-driven environmental monitoring. Thus, the implementation of EO-ML for environmental monitoring proved to be scalable, efficient, and robust. Hence, a reliable baseline for continuous environmental surveillance across complex and dynamic ecosystems can be established by combining satellite-driven insights with machine learning capabilities.

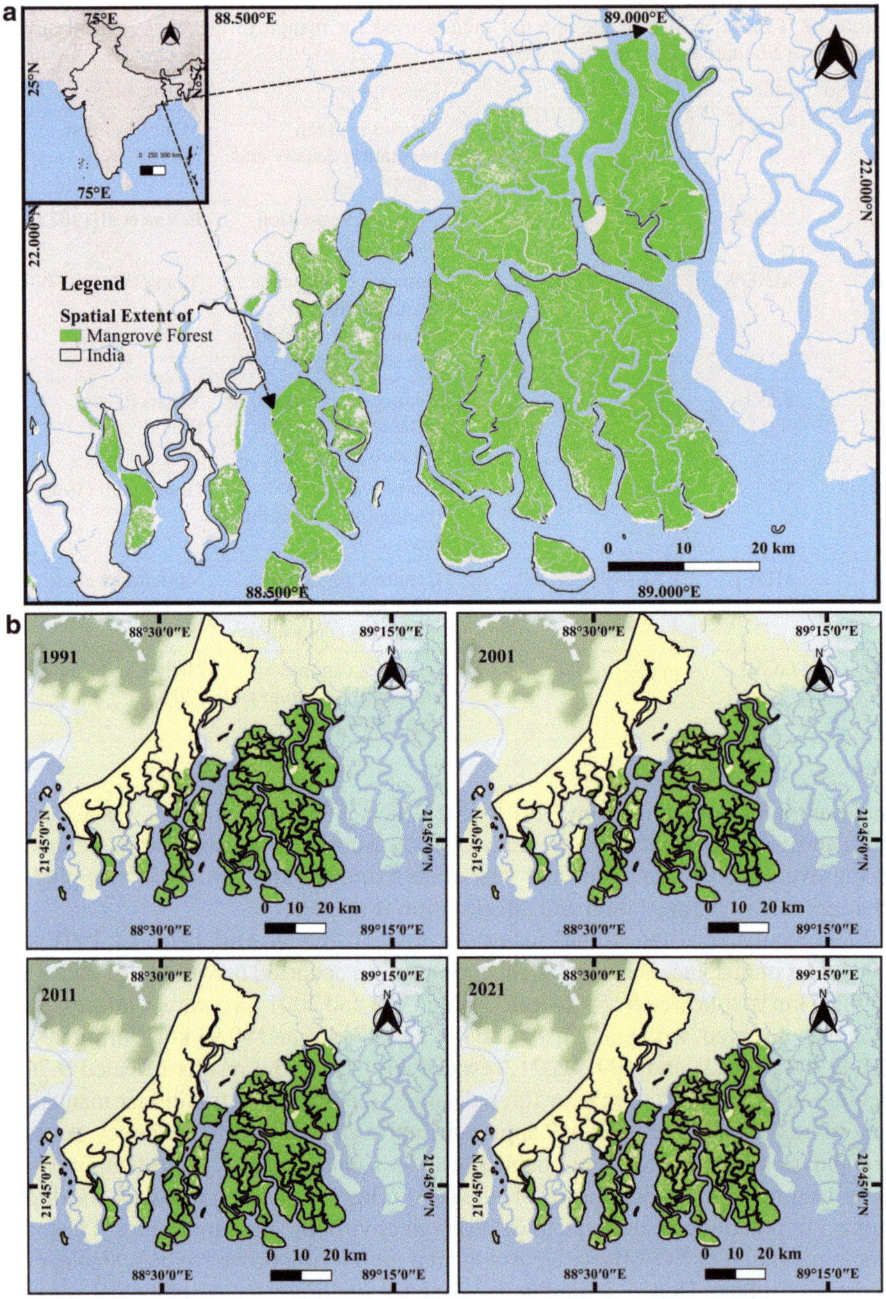


**Fig. 3.3** **a** Geographical location of the study area of Sundarbans, India; **b** Spectral indices-based RF model predicted spatial mapping of mangrove forest extent from 1991–2021; **c** Afforestation extent of mangrove forest cover from 1991–2021; **d** Deforestation extent of mangrove forest cover from 1991–2021

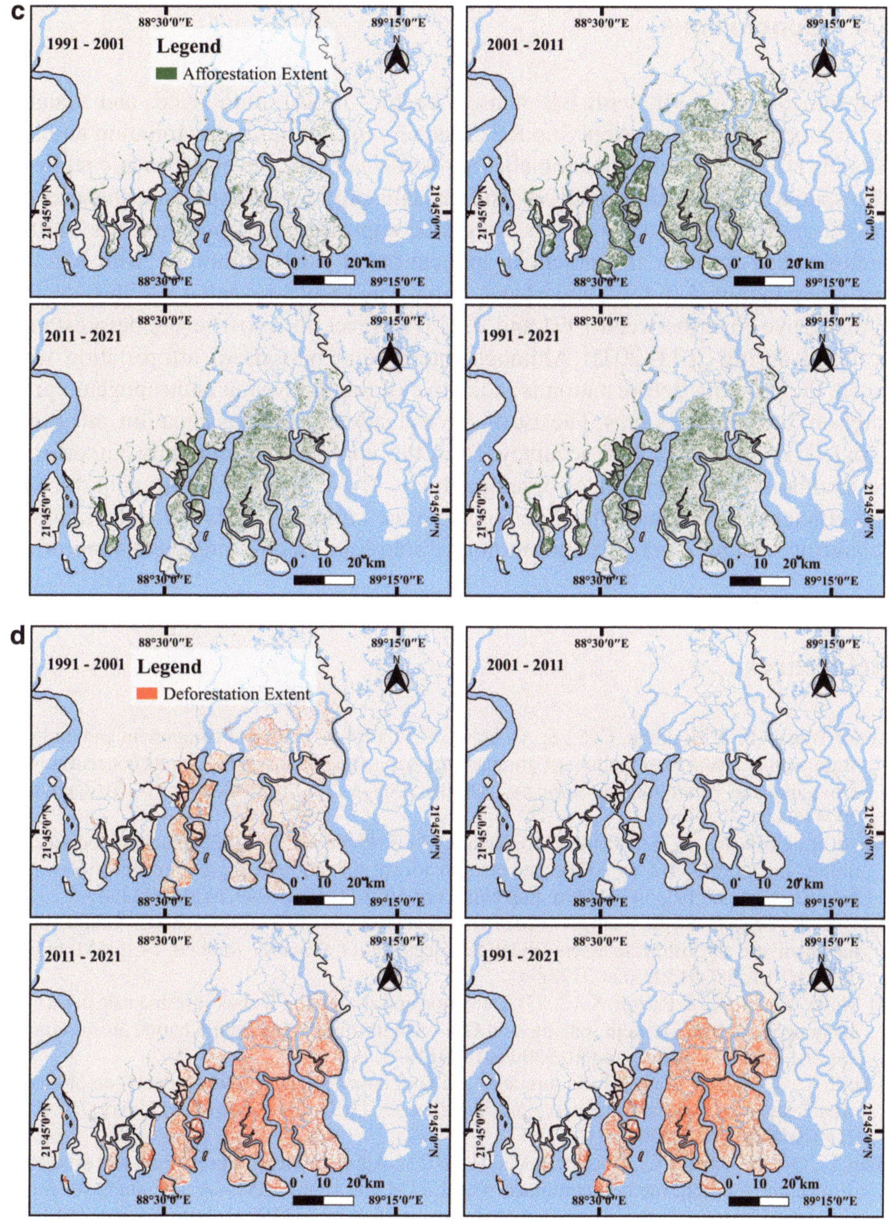


**Fig. 3.3** (continued)

## 3.4 Conclusions

The integration of ML with EO datasets enables robust, automated, and scalable environmental analysis. Applying RF classifiers to long-term precipitation and SPI datasets effectively detected anomalies across decades and diverse climatic regimes. The spatial identification of flood- and drought-prone zones further enhanced the insight-generation capacity of this integrated approach. This potential was further demonstrated through the practical application of GEE combined with ML for mangrove mapping in the Sundarbans, India. The results revealed a 9.76% decline in mangrove cover between 1991 and 2021. However, the most severe deforestation occurred during 2011–2021. Although intermittent periods of afforestation were noted, the growing deforestation is indicative of the increasing anthropogenic pressures on coastal ecosystems. The use of spectral index-based classification, integrated with RF, achieved over 90% accuracy. Thus, this approach substantially outperforms the traditional manual or pixel-based methods. These findings, along with the ease and efficiency of the workflow, demonstrate the strength of EO-ML integration in delivering scalable, accurate, and automated solutions for environmental monitoring.

## References

Abdul-Wahab, S. A., Bakheit, C. S., & Al-Alawi, S. M. (2005). Principal component and multiple regression analysis in modelling of ground-level ozone and factors affecting its concentrations. *Environmental Modelling & Software, 20*(10), 1263–1271. https://doi.org/10.1016/j.envsoft.2004.09.001

Abedi, R., Costache, R., Shafizadeh-Moghadam, H., & Pham, Q. B. (2022). Flash-flood susceptibility mapping based on XGBoost, random forest and boosted regression trees. *Geocarto International, 37*(19), 5479–5496. https://doi.org/10.1080/10106049.2020.1834258

Ajin, R. S., Segoni, S., & Fanti, R. (2024). Optimization of SVR and CatBoost models using metaheuristic algorithms to assess landslide susceptibility. *Scientific Reports, 14*, 24851. https://doi.org/10.1038/s41598-024-31788-4

Al-Mulla, Y., Ali, A., & Parimi, K. (2023). Detection and analysis of Dubas-infested date palm trees using deep learning, remote sensing, and GIS techniques in Wadi Bani Kharus. *Sustainability, 15*(19), 14045. https://doi.org/10.3390/su151914045

Anand, V., & Oinam, B. (2020). Future land use land cover prediction with special emphasis on urbanization and wetlands. *Remote Sensing Letters, 11*(3), 225–234. https://doi.org/10.1080/2150704X.2019.1703791

Bian, L., Qin, X., Zhang, C., Guo, P., & Wu, H. (2023). Application, interpretability and prediction of machine learning method combined with LSTM and LightGBM—A case study for runoff simulation in an arid area. *Journal of Hydrology, 625*, Article 130091. https://doi.org/10.1016/j.jhydrol.2023.130091

Buzna, C., Horablaga, M. N., Herbei, M. V., & Sala, F. (2022). Models for describing the dynamics of forest vegetation based on remote sensing techniques. *International Multidisciplinary Scientific GeoConference: SGEM, 22*(3.2), 361–368. https://doi.org/10.5593/sgem2022/3.2/s15.046

Cao, L., He, X., Chen, S., & Fang, L. (2023). Assessing forest quality through forest growth potential, an index based on improved CatBoost machine learning. *Sustainability, 15*(11), 8888. https://doi.org/10.3390/su15118888

Chowdhury, K., & Sarwat, M. (2022, November). GeoTorch: A spatiotemporal deep learning framework. In *Proceedings of the 30th International Conference on Advances in Geographic Information Systems* (pp. 1–4). https://doi.org/10.1145/3557915.3560992

Chowdhury, K., & Sarwat, M. (2024, May). Deep learning with spatiotemporal data: A deep dive into GeoTorchAI. In *2024 IEEE 40th International Conference on Data Engineering (ICDE)* (pp. 5156–5169). IEEE. https://doi.org/10.1109/ICDE60468.2024.00502

Coskun, H. G., Tanik, A., Alganci, U., & Cigizoglu, H. K. (2008). Determination of environmental quality of a drinking water reservoir by remote sensing, GIS and regression analysis. *Water, Air, and Soil Pollution, 194*, 275–285. https://doi.org/10.1007/s11270-008-9728-y

Dey, A., Sahoo, D. P., Kumar, R., & Remesan, R. (2022). A multimodel ensemble machine learning approach for CMIP6 climate model projections in an Indian River Basin. *International Journal of Climatology, 42*(16), 9215–9236. https://doi.org/10.1002/joc.7736

Dixit, S., & Pandey, K. K. (2024). Spatiotemporal variability identification and analysis for non-stationary climatic trends for a tropical river basin of India. *Journal of Environmental Management, 365*, Article 121692. https://doi.org/10.1016/j.jenvman.2024.121692

Dota, M. A., Cugnasca, C. E., & Barbosa, D. S. (2015). Comparative analysis of decision tree algorithms on quality of water contaminated with soil. *Ciência Rural, 45*, 267–273. https://doi.org/10.1590/0103-8478cr20130950

El Naqa, I., & Murphy, M. J. (2015). What is machine learning? In *Machine learning in radiation oncology: Theory and applications* (pp. 3–11). Springer International Publishing. https://doi.org/10.1007/978-3-319-18305-3_1

Feng, S., Li, W., Xu, J., Liang, T., Ma, X., Wang, W., & Yu, H. (2022). Land use/land cover mapping based on GEE for the monitoring of changes in ecosystem types in the upper Yellow River Basin over the Tibetan Plateau. *Remote Sensing, 14*(21), 5361. https://doi.org/10.3390/rs14215361

Fusco, T., Bi, Y., Wang, H., & Browne, F. (2020). Data mining and machine learning approaches for prediction modelling of schistosomiasis disease vectors: Epidemic disease prediction modelling. *International Journal of Machine Learning and Cybernetics, 11*(6), 1159–1178. https://doi.org/10.1007/s13042-019-01045-4

Goyal, M. K., Gupta, A. K., Das, J., Jain, V., & Rakkasagi, S. (2023). Heatwave magnitude impact over Indian cities: CMIP6 projections. *Theoretical and Applied Climatology, 154*(3), 959–971. https://doi.org/10.1007/s00704-023-04287-6

Green, E. P., Clark, C. D., Mumby, P. J., Edwards, A. J., & Ellis, A. C. (1998). Remote sensing techniques for mangrove mapping. *International Journal of Remote Sensing, 19*(5), 935–956. https://doi.org/10.1080/014311698215801

Hamed, K. H., & Rao, A. R. (1998). A modified Mann-Kendall trend test for autocorrelated data. *Journal of Hydrology, 204*(1–4), 182–196. https://doi.org/10.1016/S0022-1694(97)00125-X

Hasmadi, M., Pakhriazad, H. Z., & Shahrin, M. F. (2009). Evaluating supervised and unsupervised techniques for land cover mapping using remote sensing data. *Geografia: Malaysian Journal of Society and Space, 5*(1), 1–10.

Hickey, S. M., & Radford, B. (2022). Turning the tide on mapping marginal mangroves with multidimensional space–time remote sensing. *Remote Sensing, 14*(14), 3365. https://doi.org/10.3390/rs14143365

Hino, M., Benami, E., & Brooks, N. (2018). Machine learning for environmental monitoring. *Nature Sustainability, 1*(10), 583–588. https://doi.org/10.1038/s41893-018-0142-9

Iqbal, Z., Shahid, S., Ismail, T., Sa'adi, Z., Farooque, A., & Yaseen, Z. M. (2022). Distributed hydrological model based on machine learning algorithm: Assessment of climate change impact on floods. *Sustainability, 14*(11), 6620. https://doi.org/10.3390/su14116620

Kharade, J., Deshmukh, P., Behl, G., Nidhi, Rauf, R. F., & Gupta, K. (2023, June). Analysis on prediction of crop diseases using TensorFlow with Keras and OpenCV technique of deep learning. In *International Conference on Data Science and Big Data Analysis* (pp. 87–98). Springer Nature Singapore. https://doi.org/10.1007/978-981-99-3494-6_9

Lary, D. J., Lary, T., & Sattler, B. (2015). Using machine learning to estimate global PM2.5 for environmental health studies. *Environmental Health Insights, 9*, EHI-S15664. https://doi.org/10.4137/EHI.S15664

Lary, D. J., Zewdie, G. K., Liu, X., Wu, D., Levetin, E., Allee, R. J., & Aurin, D. (2018). Machine learning applications for Earth observation. In *Earth observation open science and innovation* (pp. 165–186). Springer. https://doi.org/10.1007/978-3-319-65633-5_8

Leggesse, E. S., Zimale, F. A., Sultan, D., Enku, T., Srinivasan, R., & Tilahun, S. A. (2023). Predicting optical water quality indicators from remote sensing using machine learning algorithms in tropical highlands of Ethiopia. *Hydrology, 10*(5), 110. https://doi.org/10.3390/hydrology10050110

Lemenkova, P., & Debeir, O. (2023). GDAL and PROJ libraries integrated with GRASS GIS for terrain modelling of the georeferenced raster image. *Technologies, 11*(2), 46. https://doi.org/10.3390/technologies11020046

Li, Q., Wang, D., Liu, F., Yu, J., & Jia, Z. (2024). LightGBM hybrid model-based DEM correction for forested areas. *PLoS ONE, 19*(10), Article e0309025. https://doi.org/10.1371/journal.pone.0309025

Meneses-Tovar, C. L. (2011). NDVI as indicator of degradation. *Unasylva, 62*(238), 39–46.

Mukonza, S. S., & Chiang, J. L. (2023). Meta-analysis of satellite observations for United Nations sustainable development goals: Exploring the potential of machine learning for water quality monitoring. *Environments, 10*(10), 170. https://doi.org/10.3390/environments10100170

Nelvidawati, N. (2024). Simple regression model analysis of the effect of temperature on rainfall in Padang City using Scikit-Learn. *International Journal of Wireless and Multimedia Communications, 1*(2), 1–6.

Nguyen, Q. H., Nguyen, H. D., Le, D. T., & Bui, Q. T. (2023). Fine-tuning LightGBM using an artificial ecosystem-based optimizer for forest fire analysis. *Forest Science, 69*(1), 73–82. https://doi.org/10.1093/forsci/fxac063

Oraegbu, A., & Jolaiya, E. (2024). Mapping soil erosion classes using remote sensing data and ensemble models. *The International Archives of the Photogrammetry, Remote Sensing and Spatial Information Sciences, 48*, 135–142. https://doi.org/10.5194/isprs-archives-XLVIII-1-2024-135-2024

Pande, C. B. (2022). Land use/land cover and change detection mapping in Rahuri watershed area (MS), India using the Google Earth Engine and machine learning approach. *Geocarto International, 37*(26), 13860–13880. https://doi.org/10.1080/10106049.2021.1956366

Persello, C., Wegner, J. D., Hänsch, R., Tuia, D., Ghamisi, P., Koeva, M., & Camps-Valls, G. (2022). Deep learning and Earth observation to support the Sustainable Development Goals: Current approaches, open challenges, and future opportunities. *IEEE Geoscience and Remote Sensing Magazine, 10*(2), 172–200. https://doi.org/10.1109/MGRS.2022.3165514

Peterson, K. T., Sagan, V., & Sloan, J. J. (2020). Deep learning-based water quality estimation and anomaly detection using Landsat-8/Sentinel-2 virtual constellation and cloud computing. *Giscience & Remote Sensing, 57*(4), 510–525. https://doi.org/10.1080/15481603.2020.1754005

Prasad, P. (2021). *Environmental geomorphology of Central West Coast of India: An integrated approach using geospatial, machine learning and ground penetrating radar techniques*. Doctoral dissertation, Goa University. Goa University Repository.

Quinn, N. W., Kumar, S., & Imen, S. (2019). Overview of remote sensing and GIS uses watershed and TMDL analyses. *Journal of Hydrologic Engineering, 24*(4), 02519002. https://doi.org/10.1061/(ASCE)HE.1943-5584.0001770

Raghavendra, R., & Kumar, N. (2023, November). Application of WEKA machine learning tools in phytoremediation for the removal of heavy metals in water treatment processes. In *2023 International Conference on Communication, Security and Artificial Intelligence (ICCSAI)* (pp. 1–5). IEEE. https://doi.org/10.1109/ICCSAI58327.2023.10232120

Raha, A., Das, S., Banerjee, K., & Mitra, A. (2012). Climate change impacts on Indian Sunderbans: A time series analysis (1924–2008). *Biodiversity and Conservation, 21*, 1289–1307. https://doi.org/10.1007/s10531-012-0249-9

Rahman, S. Z., Senthil, R., Ramalingam, V., & Gopal, R. (2023). Predicting infectious disease outbreaks with machine learning and epidemiological data. *Journal of Advanced Zoology, 44*(S4), 110–121.

Rakkasagi, S., & Goyal, M. K. (2025). Are rapid onset drying events escalating forest fires across India's ecoregions? *Environmental Research Letters.*

Rakkasagi, S., Edris, S. G., & Goyal, M. K. (2025). A socio-technical framework for assessing flash drought risks to key biodiversity areas: A scientists' warning. *Technology in Society, 82*, Article 102893.

Rakkasagi, S., Goyal, M. K., & Jha, S. (2024). Evaluating the future risk of coastal Ramsar wetlands in India to extreme rainfalls using fuzzy logic. *Journal of Hydrology, 632*, Article 130869.

Raschka, S., Liu, Y. H., & Mirjalili, V. (2022). *Machine learning with PyTorch and Scikit-Learn: Develop machine learning and deep learning models with Python.* Packt Publishing Ltd.

Rautela, K. S., Singh, S., & Goyal, M. K. (2024a). Characterizing the spatio-temporal distribution, detection, and prediction of aerosol atmospheric rivers on a global scale. *Journal of Environmental Management, 351*, Article 119675.

Rautela, K. S., Singh, S., & Goyal, M. K. (2024b). Resilience to air pollution: A novel approach for detecting and predicting aerosol atmospheric rivers within earth system boundaries. *Earth Systems and Environment*, 1–23.

Ravindiran, G., Rajamanickam, S., Kanagarathinam, K., Hayder, G., Janardhan, G., Arunkumar, P., & Muniasamy, S. K. (2023). Impact of air pollutants on climate change and prediction of air quality index using machine learning models. *Environmental Research, 239*, Article 117354. https://doi.org/10.1016/j.envres.2023.117354

Reiser, P., Eberhard, A., & Friederich, P. (2021). Graph neural networks in TensorFlow-Keras with RaggedTensor representation (kgcnn). *Software Impacts, 9*, Article 100095. https://doi.org/10.1016/j.simpa.2021.100095

Rigatti, S. J. (2017). Random forest. *Journal of Insurance Medicine, 47*(1), 31–39. https://doi.org/10.17849/insm-47-1-31-39.1

Roy, S. (2024). Decoding the declining mangrove ecosystems: An analysis of carbon dynamics, salinity changes, and deforestation using remote sensing and machine learning. *AGU Fall Meeting Abstracts.*

Roy, S., & Chintalacheruvu, M. R. (2023a, February). LULC dynamics study and modeling of urban land expansion using CA-ANN. In *International Conference on Science, Technology and Engineering* (pp. 79–90). Singapore: Springer Nature Singapore.

Roy, S., & Chintalacheruvu, M. R. (2023b). Assessing long-term changes in the lower Godavari River Basin using geo-spatial technology: An analysis of river bank erosion and accretion. *Sustainable Lifestyle Approaches Towards Environment and Climate Change, EIACP (PC-RP) Special*, 36–49.

Roy, S., & Chintalacheruvu, M. R. (2024). Delineating hydro-geologically constrained groundwater zones in the Himalayan River basins of India through an innovative ensemble of hypsometric analysis and machine learning algorithms. *Earth Science Informatics, 17*(1), 501–526. https://doi.org/10.1007/s12145-023-01127-6

Roy, S., & Goyal, M. (2025). Climate variability and extremes in Indian River Basins: Trends, regional disparities, and urban risk assessment (No. EGU25-1114). In *Copernicus Meetings.*

Sagan, V., Coral, R., Bhadra, S., Alifu, H., Al Akkad, O., Giri, A., & Esposito, F. (2024). Hyperfidelis: A software toolkit to empower precision agriculture with GeoAI. *Remote Sensing, 16*(9), 1584. https://doi.org/10.3390/rs16091584

Sawant, S., Garg, R. D., Meshram, V., & Mistry, S. (2023). Sen-2 LULC: Land use land cover dataset for deep learning approaches. *Data in Brief, 51*, Article 109724. https://doi.org/10.1016/j.dib.2023.109724

Şen, Z. (2012). Innovative trend analysis methodology. *Journal of Hydrologic Engineering, 17*(9), 1042–1046. https://doi.org/10.1061/(ASCE)HE.1943-5584.0000556

Singh, S., Gururani, D. M., Kumar, A., Kumar, Y., Bohra, M. S., & Mehta, P. (2023). Declining groundwater level and its impact on irrigation and agro-production. In *Advancement of GI-science and sustainable agriculture: A multi-dimensional approach* (pp. 217–224). Cham: Springer Nature Switzerland.

Tassi, A., & Vizzari, M. (2020). Object-oriented LULC classification in Google Earth Engine combining SNIC, GLCM, and machine learning algorithms. *Remote Sensing, 12*(22), 3776. https://doi.org/10.3390/rs12223776

Valderrama-Landeros, L., Flores-de-Santiago, F., Kovacs, J. M., & Flores-Verdugo, F. (2018). An assessment of commonly employed satellite-based remote sensors for mapping mangrove species in Mexico using an NDVI-based classification scheme. *Environmental Monitoring and Assessment, 190*, 1–13. https://doi.org/10.1007/s10661-017-6403-2

Venancio, L. P., Mantovani, E. C., do Amaral, C. H., Neale, C. M. U., Gonçalves, I. Z., Filgueiras, R., & Campos, I. (2019). Forecasting corn yield at the farm level in Brazil based on the FAO-66 approach and soil-adjusted vegetation index (SAVI). *Agricultural Water Management, 225*, 105779. https://doi.org/10.1016/j.agwat.2019.105779

Wang, F., Wang, X., & Li, S. (2025). Explainable machine learning for predictive modeling of blowing snow detection and meteorological feature assessment using XGBoost-SHAP. *PLoS ONE, 20*(3), Article e0318835. https://doi.org/10.1371/journal.pone.0318835

Wang, M., Li, Y., Yuan, H., Zhou, S., Wang, Y., Ikram, R. M. A., & Li, J. (2023). An XGBoost-SHAP approach to quantifying morphological impact on urban flooding susceptibility. *Ecological Indicators, 156*, Article 111137. https://doi.org/10.1016/j.ecolind.2023.111137

Yang, S., Yang, D., Chen, J., Santisirisomboon, J., Lu, W., & Zhao, B. (2020). A physical process and machine learning combined hydrological model for daily streamflow simulations of large watersheds with limited observation data. *Journal of Hydrology, 590*, Article 125206. https://doi.org/10.1016/j.jhydrol.2020.125206

Yang, X., Zhao, S., Qin, X., Zhao, N., & Liang, L. (2017). Mapping of urban surface water bodies from Sentinel-2 MSI imagery at 10 m resolution via NDWI-based image sharpening. *Remote Sensing, 9*(6), 596. https://doi.org/10.3390/rs9060596

Zhou, Z.-H. (2021). *Machine learning*. Springer Nature. https://doi.org/10.1007/978-981-15-3414-5

# Chapter 4
# Case Study of the Narmada River Basin

## 4.1 Introduction

To better understand the various aspects of environmental monitoring of the Narmada River Basin, key factors such as physiographic conditions, slope characteristics, land use and land cover (LULC) changes, and spatial and temporal variations in weather were assessed. By employing an integrated approach that combined earth observation data (EO) with machine learning (ML) techniques, these factors were effectively analyzed and monitored, providing deeper insights into the evolving landscape of the Narmada River Basin (Fig. 4.1).

## 4.2 Environmental Overview of the Narmada River Basin

The Narmada River Basin, historically known as the "Rewa River," extends between longitudes 72° 32′ E–81° 45′ E and latitudes 21° 20′ N–23° 45′ N, covering approximately 98,796 km$^2$. Revered as the "daughter of Shiva," the Narmada holds a sacred place in Hindu mythology, with its origins described in ancient scriptures such as the Mahabharata, Ramayana, and Matsya Purana. This basin has served as a cradle of ancient civilizations, fostering trade, commerce, and cultural exchange (NCA, 2025). However, beyond its religious and cultural significance, the Narmada River Basin is also a vital water resource for the peninsular India (Sharma et al., 2008; Singh & Kumar, 2024; Singh et al., 2024). It originates from the Amarkantak Plateau in the Anuppur district of Madhya Pradesh, India, and travels westward for 1,357 km before emptying into the Arabian Sea at the Gulf of Khambhat, near Bharuch, Gujarat. The long journey traversed by the Narmada River is a witness to a wide variety of physiographic, climatic, demographic, and overall environmental diversities. However, with the advancements in Earth Observation and machine learning, these environmental diversities of the entire 1,357 km stretch traversed by the river can be minutely

 
S. Roy and M. K. Goyal, *Environmental Monitoring with Integrated Earth Observation Data and Machine Learning*, Studies in Computational Intelligence 1278,
https://doi.org/10.1007/978-3-032-19935-5_4

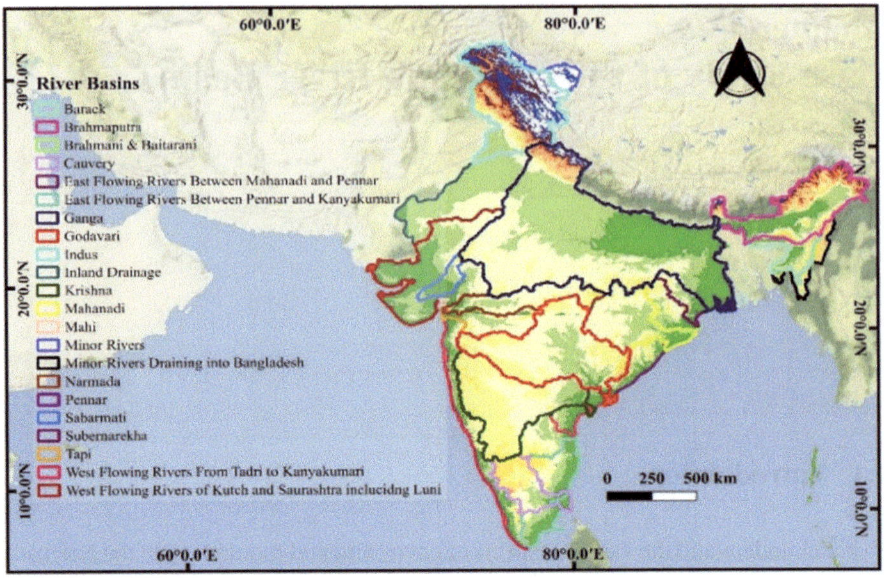


**Fig. 4.1** Map highlighting the geographical locations and spatial extents of the Indian River Basins

analyzed. In this context, EO topographic datasets, particularly digital elevation models (DEMs), are instrumental in studying land elevation and terrain features. These datasets are derived from satellite missions that capture detailed elevation data of landforms. Hence, a comprehensive tabulation of available EO topographic datasets capturing elevation data is given in Table 4.1 (Fig. 4.2).

Beyond these single-source datasets, EO datasets are often integrated to enhance accuracy, spatial coverage, and usability. Thus, by combining multiple topographic datasets (Table 4.2), the limitations of single-source EO data can be overcome, improving comprehensive environmental monitoring. This integration of EO datasets also enables achieving better resolution and increases precision in geospatial studies.

From the wide range of available DEMs (Tables 4.1 and 4.2), the selection of a DEM depends on the study's purpose and specific objectives (Chang & Tsai, 1991; Roy et al., 2025; Saksena & Merwade, 2015). Key factors such as spatial resolution, vertical accuracy, data availability, and processing requirements play a crucial role in determining the most suitable dataset. For large-scale studies like regional hydrological modeling and terrain classification, moderate-resolution DEMs such as SRTM or ASTER GDEM provide sufficient accuracy while maintaining computational efficiency. However, for high-precision applications such as flood modeling, landslide susceptibility analysis, and infrastructure planning, higher-resolution datasets like CartoDEM (2.5 m) and TanDEM-X (12 m) offer greater detail and accuracy in capturing small-scale topographic variations. Among these datasets, CartoDEM, derived from Cartosat-1 stereo imagery, is available in multiple resolutions, including a higher-resolution 2.5 m DEM and a 30 m version for broader applications. The

**Table 4.1** Earth Observation topographic datasets

| Dataset | Satellite/mission | Country of origin | Source | Spatial resolution | Temporal resolution |
|---|---|---|---|---|---|
| SRTM DEM | Shuttle Radar Topography Mission (SRTM) | USA | NASA | 30 m (Global) | Single mission (2000) |
| CartoDEM | Cartosat-1 | India | ISRO | 2.5–30 m | Varies (Stereo imaging from 2005) |
| ALOS World 3D (AW3D30) | Advanced Land Observing Satellite (ALOS) | Japan | JAXA | 30 m (Global), 5 m (Commercial) | 2006–2011 (ALOS-1), 2014–2021 (ALOS-2) |
| ASTER GDEM | Advanced Spaceborne Thermal Emission and Reflection Radiometer (ASTER) | USA/Japan | NASA | 30 m | 1999-present (on-demand acquisition) |
| TanDEM-X DEM | TerraSAR-X add-on for Digital Elevation Measurement (TanDEM-X) | Germany | DLR | 12 m (Global), 90 m (Public) | 2010–2014 (Acquired in two phases) |

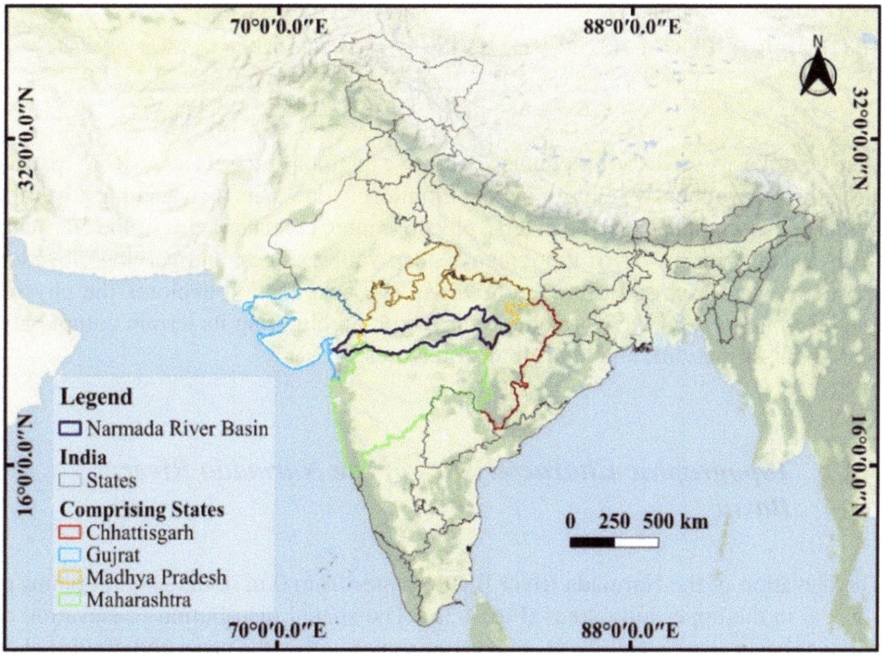


**Fig. 4.2** Geographical location of the Narmada River Basin highlighting the states it traverses

**Table 4.2** EO topographic datasets from combined missions

| Dataset | Sources used | Country of origin | Source | Spatial resolution |
|---|---|---|---|---|
| ETOPO1 | SRTM, ICESat, GEBCO, NOAA bathymetric data | USA | NOAA | 1 arc-minute |
| GTOPO30 | SRTM, cartographic maps, photogrammetry | USA | USGS | 30 arc-seconds |
| Earth2014 | SRTM, ICESat, airborne LiDAR data | Germany/Australia | TUM | 1 arc-minute |
| SRTM30_PLUS | SRTM DEM, bathymetric data, interpolation | USA | SIO | 30 arc-seconds |
| Copernicus DEM | Sentinel-1 SAR, TanDEM-X data | Europe | ESA | 30 m (Global), 90 m (Public) |
| ArcticDEM | Various commercial satellites | USA | PGC | 2 m |
| FABDEM | Multiple EO DEMs, vegetation-corrected | UK | FABDEM | 30 m |
| EarthDEM | Various commercial satellites | USA | PGC | 2 m |
| GMTED2010 | SRTM, ASTER, topographic maps | USA | USGS | 7.5 arc-seconds |

30 m CartoDEM serves as a valuable alternative to global datasets like SRTM, particularly for comprehensive environmental monitoring due to its high accuracy over the Indian subcontinent. Thus, the diverse physiographic characteristics of the Narmada River Basin, with its varied topography, steep valley slopes, and varying elevation gradients, were assessed using CartoDEM. The next section explores the physiographic conditions of the Narmada River Basin, highlighting its terrain complexity, geological formations, and hydrological features.

### *4.2.1 Topographic Characteristics of the Narmada River Basin*

The elevation of the Narmada River Basin ranges from 0 m in the Gujarat plains to 1261 m in the upper hilly areas (Fig. 4.3a). The spatial distribution of elevation of the river basin also reveals distinct elevation zones within the basin and provides key insights into its geological and erosional stages. The mean elevation of the basin is 355 m, with the highest point recorded at 1231 m. A significant portion (27.26%) of the basin falls within the dominant mid-altitude elevation range of 200–300 m.

The elevation distribution map (Fig. 4.3a) illustrates the spatial variability of terrain height across the spatial extent of basin. The river basin also exhibits a relatively even elevation distribution, with fewer areas in low-lying zones (<10 m) or very high-altitude regions (>1,000 m). The elevation characteristics of the Narmada Basin are further given in Table 4.3.

Additionally, the hypsometric curve which plots relative elevation against relative area was analyzed to assess the basin's geological stage of erosion (Pérez-Peña et al., 2009; Roy & Chintalacheruvu, 2024a, 2024b; Strahler, 1956). The hypsometric curve (Fig. 4.3b) of the river basin exhibits a slightly concave-upward and convex-downward shape, suggesting that the basin is in a late young to early equilibrium or mature geological stage. This indicates that erosion is more dominant at higher elevations, particularly on steep slopes, where processes like mass soil movement

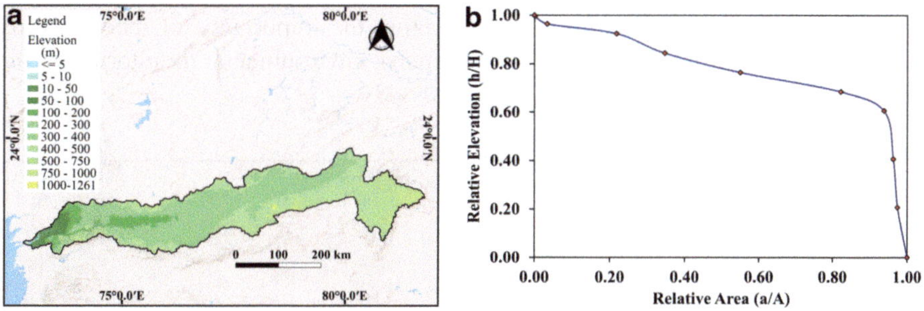


**Fig. 4.3** **a** Spatial distribution of elevation across the Narmada River Basin; **b** Distribution of relative elevation versus relative area of the Narmada River Basin

**Table 4.3** Elevation zones of the Narmada Basin

| Sl. No | Elevation (m) | Area (km$^2$) | % of Total area |
|---|---|---|---|
| 1 | <5 | 2325.72 | 2.42 |
| 2 | 5–10 | 145.22 | 0.15 |
| 3 | 10–50 | 1077.61 | 1.12 |
| 4 | 50–100 | 2325.01 | 2.42 |
| 5 | 100–200 | 11,208.40 | 11.68 |
| 6 | 200–300 | 26,154.60 | 27.26 |
| 7 | 300–400 | 19,481.90 | 20.30 |
| 8 | 400–500 | 12,335.70 | 12.86 |
| 9 | 500–750 | 17,654.30 | 18.40 |
| 10 | 750–1000 | 3114.34 | 3.25 |
| 11 | 1000–1261 | 136.40 | 0.14 |
| Highest Elevation: 1261 m | | | |

and accelerated erosion occur (Harlin, 1978; Pérez-Peña et al., 2009; Roy & Chintalacheruvu, 2023; Roy, 2024; Rakkasagi & Goyal, 2025; Rakkasagi et al., 2025; Strahler, 1952; Shekar & Mathew, 2022).

Thus, the utilization of Earth Observation topographic datasets, such as CartoDEM, enables a detailed analysis of the topographic characteristics and geological stage of a river basin. Additionally, the DEM of the basin can be further utilized to assess the spatial distribution of slope, providing critical insights into terrain variations. The slope map of the Narmada River Basin (Fig. 4.4) reveals terrain ranging from nearly level to very steep slopes, with most of the basin exhibiting moderate elevation (200–300 m) and relatively even slope variations. However, while the central and northern portions of the basin consist predominantly of gently sloping landforms, the southern portions feature significantly steep slopes, particularly in regions where the terrain transitions into the Satpura range. Moreover, these steep slopes in the elevated peripheral areas pose challenges related to soil erosion, landslides, and runoff acceleration, emphasizing the importance of incorporating additional EO datasets for a more comprehensive environmental monitoring of the Narmada River Basin.

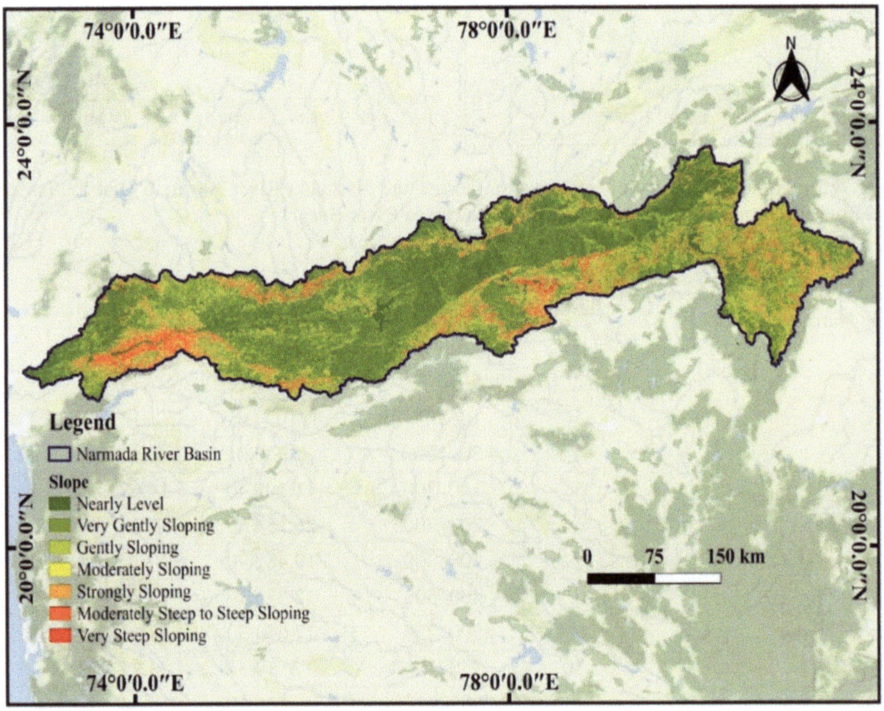


**Fig. 4.4** Spatial distribution of slope across the Narmada River Basin

### 4.2.2 *Stream Network Characteristics of the Narmada River Basin*

Moreover, multiple EO datasets can be further integrated to provide crucial insights into river networks, watershed boundaries, and drainage patterns. With advancements in geospatial platforms like Google Earth Engine (GEE), a cloud-based computational platform hosting multi-petabyte open-access EO datasets processing and integrating multiple large-scale environmental data has become more convenient (Roy & Chinatalacheruvu, 2023; Roy & Chintalacheruvu, 2024b). Although EO datasets like free-flowing rivers (Lehner et al. 2008) offer valuable information on global river connectivity, ensuring consistency between the derived stream network and the DEM resolution is essential. To address this, ArcMap 10.3 was used to perform stream network delineation from CartoDEM and assess stream characteristics within the Narmada River Basin (Fig. 4.5).

The Narmada River spans 1357 km, with major tributaries like Shakkar, Dudhi, Tawa, and Orsang joining from both sides. Topographic and stream network analysis of the river basin provided insights into the basin's geometry, drainage, and relief parameters (Table 4.4), revealing its hydrological response, susceptibility to environmental hazards, and overall stability (Roy & Chintalacheruvu, 2024a).

The basin drainage parameter of drainage density ($D_d$) of 0.27 km/km$^2$ suggests a low to moderate drainage network and highlights its greater infiltration capacity and

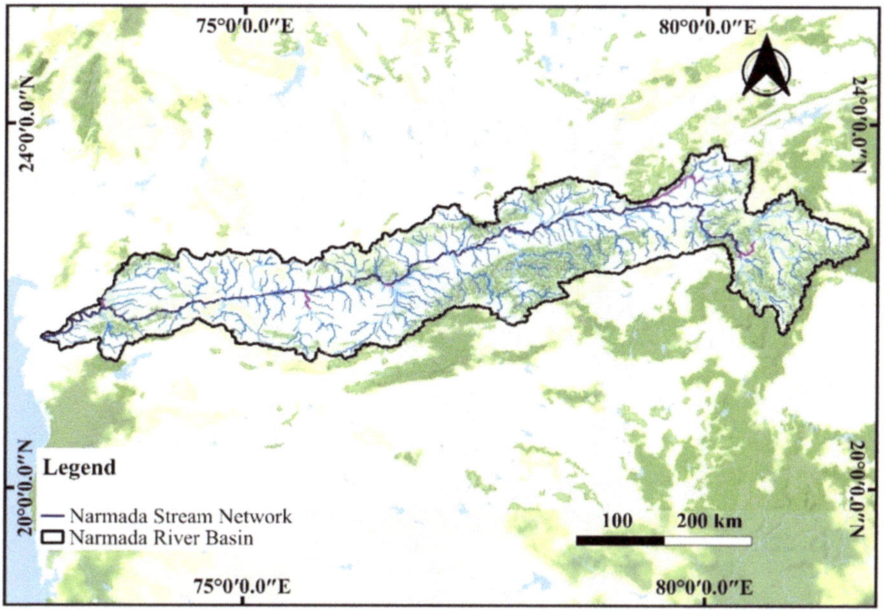


Fig. 4.5 Narmada stream network map

**Table 4.4** Drainage, geometry, and relief characteristics of the Narmada River Basin

| Sl. No | Drainage parameters | | | | | | | | | |
|---|---|---|---|---|---|---|---|---|---|---|
| 1 | Stream number ($N_u$) | Stream length ($L_u$) | Drainage density ($D_d$) (Km/Km$^2$) | Dissection index ($D_i$) | Length of overland flow ($L_o$) (km) | Texture ratio ($T_r$) | | | | |
| | 25,677 | 268,822 | 0.27 | 1.01 | 1.87 | 1.65 | | | | |
| 2 | Basin geometry parameters | | | | | | | | | |
| | Circularity ratio ($R_c$) | Compactness coefficient ($C_C$) | Constant of channel maintenance ($C_m$) | Ellipticity ratio ($E_r$) | Elongation ratio ($E_L$) | Form factor ($F_f$) | Shape factor ($S_f$) | Watershed area (A) km$^2$ | Watershed length ($W_l$) | Watershed perimeter ($W_p$) |
| | 0.08 | 3.50 | 3.73 | 6.43 | 0.22 | 0.12 | 8.19 | 98,796 | 885.98 | 3834.52 |
| 3 | Basin relief parameters | | | | | | | | | |
| | Basin relief (H) (m) | Mean slope ($S_o$) (°) | Relative relief ($R_r$) | Relief ratio ($R_{rr}$) | Ruggedness number (Rn) | | | | | |
| | 1261 | 5.10 | 35.15 | 1.52 | 0.07 | | | | | |

lower surface runoff. Thus, it reflects the basins enhanced groundwater recharge and reduced flash flood frequency susceptibility (Strahler, 1957). However, areas with higher stream frequency ($F_s$) and steeper slopes may still experience localized rapid runoff and flash floods during intense rainfall events (Şener & Arslanoğlu, 2023). Moreover, the dissection index ($D_i$) of 1.01 also confirms active erosion, particularly in the upper Narmada Basin with higher relief (Schumm, 1963). This is also further reinforced by the low length of overland flow ($L_o$) of 1.87 km in the Narmada River Basin. Lower Lo accelerates sediment transport and soil erosion in steeper terrains of the basin (Horton, 1945).

The basin geometry parameter of elongation ratio ($E_L$) of 0.22 and circularity ratio ($R_c$) of 0.08 highlight the basin's elongated shape. This elongated shape of the basin results in delayed peak discharge and reduces flood susceptibility compared to circular basins with faster hydrological responses (Miller, 1953). Further, the compactness coefficient ($C_C$) of 3.50 also indicates prolonged flood peaks and makes reservoir regulation an effective flood mitigation strategy within the Narmada River Basin (Smith, 1950). The basin relief (H) of 1,261 m and relative relief ($R_r$) of 35.15 indicate significant elevation variations, particularly in the upper and middle regions. The mean slope ($S_o$) of 5.10° also contributes to accelerated runoff, high sediment yield, and erosion risk, especially in deforested areas lower regions of the river basin (Ray & Goel, 2019). Although the ruggedness number ($R_n$) of 0.07 is relatively low, it still suggests topographic irregularities affecting drainage, erosion, and landslide risks (Melton, 1957; Roy & Chintalacheruvu, 2024a, 2024b).

Thus, the low drainage density and moderate relief of the Narmada River Basin enhance water retention and supports sustainable irrigation and groundwater recharge (Toma et al., 2023). Although the basin's elongated shape and low drainage density reduce flash flood risks, the steep terrain and active dissection in the upper reaches of the river contribute to the rapid sediment transport and increased siltation in downstream reservoirs like Sardar Sarovar (Parasuraman et al., 2010). However, to effectively assess environmental hazards and devise mitigation strategies, it is imperative to assess the spatial distribution of LULC, other anthropogenic influences such as infrastructure, and existing climatic conditions across the Narmada River Basin. In this pursuit, several open-source EO datasets were utilized and extracted for further evaluation of the basin's overall environmental conditions.

### *4.2.3 Climatic Characteristics and Climatic Variation Assessment Using EO Climatic Datasets*

Earth Observation climatic datasets offer comprehensive and consistent global data and enable detailed analysis of different climatic parameters. These datasets are derived from satellite missions and provide valuable insights into precipitation patterns, temperature variations, humidity levels, and other atmospheric conditions pertinent to river basin studies (Dubey et al., 2021). Thus, a comprehensive tabulation of some of the most used global EO climatic datasets is given in Table 4.5.

**Table 4.5** Earth observation climatic datasets and their sources

| Dataset name | Source | Satellite program | Citations |
|---|---|---|---|
| Tropical Rainfall Measuring Mission (TRMM) | NASA and JAXA | TRMM satellite | Kummerow et al. (1998) |
| Global Precipitation Measurement (GPM) | NASA and JAXA | GPM Core observatory | Huffman et al. (2015) |
| Moderate Resolution Imaging Spectroradiometer (MODIS) Land Surface Temperature (LST) | NASA | Terra and aqua satellites | Wan et al. (2015) |
| ERA5 Reanalysis | European Centre for Medium-Range Weather Forecasts (ECMWF) | Various satellites and in-situ observations | Hersbach et al. (2020) |
| Climate Hazards Group InfraRed Precipitation with Station data (CHIRPS) | Climate Hazards Center, University of California, Santa Barbara | Multiple satellites and ground stations | Funk et al. (2015) |

Among the above-mentioned EO climatic datasets (Table 4.5), the ERA5 dataset (Spatial Resolution of 11132 m), sourced from the European Centre for Medium-Range Weather Forecasts (ECMWF), was used for the climatic characteristics analysis of the Narmada River Basin. The ERA5 EO climatic dataset also includes additional parameters like surface pressure, wind speed, soil temperature, humidity, etc., and can further enhance the understanding of the river basin's climatic characteristics by providing a more detailed and refined analysis.

Thus, using the ERA5 dataset, the mean total precipitation sum during this period was observed to be 95.66 mm (Fig. 4.6a). Notably, the years 2010 and 2020 experienced higher precipitation, especially in the upper basin regions. Similarly, the temperature analysis indicates that the middle and lower sections of the basin consistently exhibited higher mean temperatures throughout the study period. The year 2010 stood out, with the central basin recording mean temperatures exceeding 299.67 Kelvin. Further, between 2000 and 2010, there was a significant temperature increase of 0.25%. Although the upper basin remained relatively cooler, with temperatures below 297.94 Kelvin, a discernible upward trend in temperature was observed over time in the Narmada River Basin. Moreover, seasonal analysis using ERA5 monthly data indicates an 18.99% rise in annual mean precipitation from 1980 to 2020. Table 4.6 also presents that 2010 and 2020 were notably wetter, with annual mean precipitation sum increases of 10.50% and 18.99%, respectively, compared to 1980. Conversely, a decline in precipitation was observed around 2000. For each selected period, data from the preceding and following years were included to mitigate the influence of anomalous years. Temperature characteristics reveal an overall decrease since 1980, except for 2010, which experienced increases across all seasons: monsoon

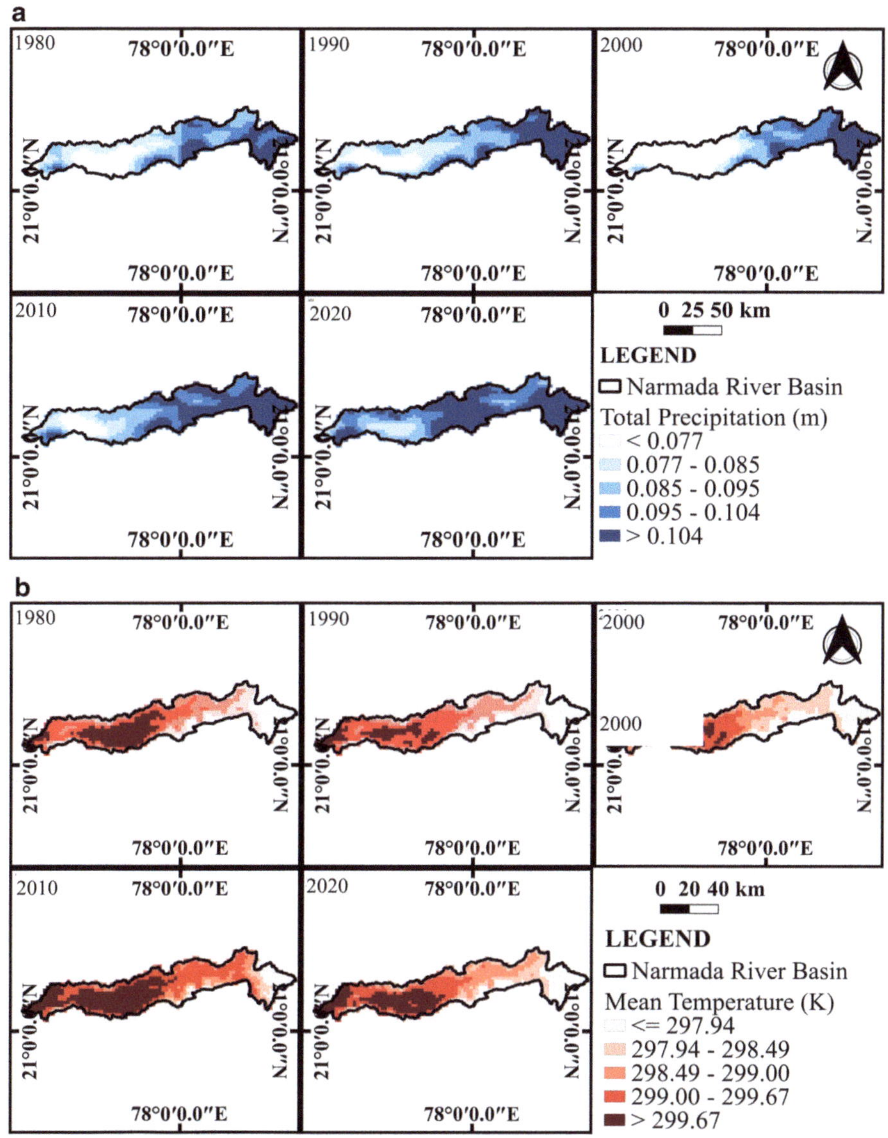


**Fig. 4.6** **a** Spatial distribution of total precipitation sum for the period 1980–2020; **b** Spatial distribution of mean temperature for the period 1980–2020

(0.05%), post-monsoon (0.08%), summer (0.27%), winter (0.27%), and annually (0.15%). Table 4.6 also presents a consistent decline in winter precipitation, with reductions of 70.61% (1980–1990), 47.04% (1980–2000), 76.50% (1980–2010), and 29.10% (1980–2020).

**Table 4.6** Yearly and seasonal variation of mean total precipitation sum and mean temperature

| | 1980–1990 (%) | 1980–2000 (%) | 1980–2010 (%) | 1980–2020 (%) |
|---|---|---|---|---|
| *Percentage change in mean total precipitation sum* | | | | |
| Mean Yearly | 4.79 | −4.20 | 10.50 | 18.99 |
| Monsoon | 6.91 | −7.05 | 10.23 | 20.17 |
| Post-monsoon | −34.10 | 47.89 | 98.83 | 24.15 |
| Summer | 124.32 | 105.99 | −48.46 | 24.96 |
| Winter | −70.61 | −47.04 | −76.50 | −29.10 |
| *Percentage change in mean temperature* | | | | |
| Mean Yearly | −0.055 | −0.104 | 0.152 | −0.005 |
| Monsoon | −0.184 | −0.206 | 0.050 | −0.056 |
| Post-monsoon | 0.014 | −0.068 | 0.081 | −0.007 |
| Summer | −0.072 | −0.034 | 0.275 | 0.071 |
| Winter | 0.126 | −0.056 | 0.275 | −0.019 |

In contrast, monsoon and post-monsoon precipitation, along with temperatures, increased in the latter two decades. The most significant temperature rise of 0.275% occurred between 1980 and 2010 during summer and winter, followed by slight decreases in 2020. These trends suggest that climate mitigation efforts may have influenced the observed changes in precipitation and temperature patterns in the last decade.

In order to further understand the interplay between precipitation and temperature, bivariate plots were generated (Fig. 4.7), delineating zones of low precipitation-high temperature, high precipitation-low temperature, low precipitation-low temperature, and high precipitation-high temperature.

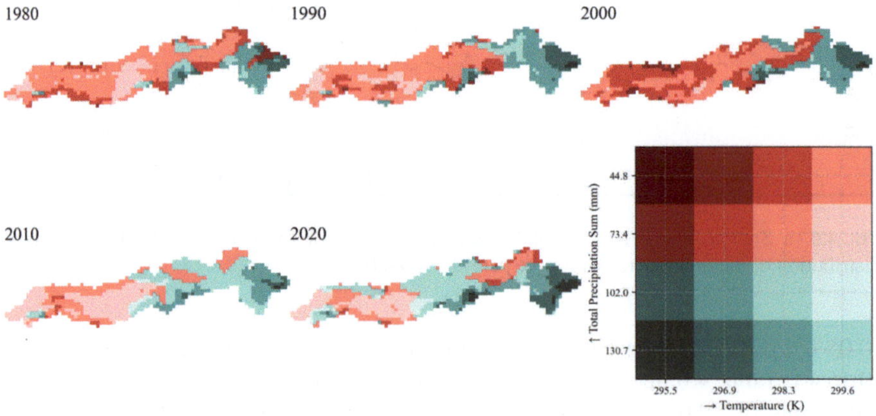


**Fig. 4.7** Temporal bivariate maps of total precipitation (mm) and mean temperature (Kelvin) for the Narmada River Basin

Figure 4.7 shows that the Upper Narmada River Basin, where the river originates, exhibits low temperature and high precipitation characteristics. Conversely, the Lower Narmada River Basin, near its mouth, is characterized by high temperature and low precipitation. In the middle portion of the river basin, a distinct north–south climatic variation is evident, where the northern half predominantly exhibits high temperature and low precipitation conditions. Moreover, as discussed in Table 4.6 and Fig. 4.6a, b, the temporal variation in these bivariate maps highlights the spatial transition of mean total precipitation sum and mean temperature over time. Specifically, the years 2010 and 2020 indicate a larger portion of the Upper Narmada Basin experiencing low temperatures with high precipitation, reinforcing the observed climate trends. Meanwhile, the Lower Narmada Basin has been increasingly facing rising temperatures and declining precipitation, suggesting possible influences from land use changes, increasing urbanization, and climatic shifts. Therefore, a detailed assessment using available EO datasets is necessary to understand these climatic variations and their broader environmental implications. Furthermore, changes in LULC patterns play a critical role in influencing regional climate and hydrological dynamics. Hence, analyzing the LULC characteristics of a region provides further valuable insights into how human activities and natural processes shape the landscape over time.

## 4.3 Integration of Earth Observation and Machine Learning for Environmental Monitoring of the Narmada River Basin

### *4.3.1 Land Use and Land Cover Characteristics*

EO satellites such as Landsat, MODIS, Sentinel, and ASTER play a crucial role in capturing land use and land cover information. These satellites are equipped with remote sensing sensors that detect electromagnetic radiation either reflected or emitted from the Earth's surface. The sensors capture data across multiple spectral bands like the visible (red, green, and blue), near-infrared (NIR), shortwave infrared (SWIR), and thermal infrared (TIR) wavelengths and enable detailed analysis of various surface features (Giri et al., 2016; Lin & Doyog, 2023). Further, the Earth's surface in satellite imagery is divided into a grid, where each grid cell (pixel) represents a specific ground area. The size of each pixel depends on the satellite's spatial resolution. For example, a spatial resolution of 10 m (Sentinel-2) means that each pixel corresponds to a 10 m $\times$ 10 m area on the ground, while coarser resolutions such as MODIS (250 m–1 km) capture larger surface areas per pixel. Thus, the radiation intensity recorded in each spectral band is stored as digital values for each pixel and results in the regeneration of a raw satellite image (Egorov, et al., 2019; Jain et al., 2023; Roy & Chintalacheruvu, 2023, 2025; Thakur et al., 2021; Zanaga

**Table 4.7** List of EO satellites for LULC mapping

| Satellite | Spatial resolution | Temporal resolution | Source |
|---|---|---|---|
| Landsat | 15 m (panchromatic), 30 m (multi-spectral) | 16 days | USGS |
| MODIS | 250 m to 1 km | Daily | NASA |
| Sentinel-2 | 10–60 m | 5 days | ESA |
| Sentinel-1 | 5 m (SAR imagery) | 6–12 days | ESA |
| Envisat (MERIS) | 300 m | 3 days | ESA |
| Proba-V | 100–300 m | Daily | ESA |
| PlanetScope | 3 m | Daily | Planet labs |
| ASTER | 15 m (VNIR), 30 m (SWIR), 90 m (TIR) | On-demand | NASA/JAXA |

et al., 2022). A consolidated list of available EO satellites used for LULC is given in Table 4.7.

Further, these raw images can be captured in true-color (raw color) or false-color formats. Raw-color images are also known as true-color images, and they display the Earth's surface as seen by the human eye using the visible spectrum (red, green, and blue bands). In contrast, false-color images use non-visible spectral bands, such as near-infrared (NIR), SWIR, or thermal infrared, to highlight specific features of the earth (Jensen & Lulla, 1987). For instance, near-infrared (NIR) and red bands are often combined in Normalized Difference Vegetation Index (NDVI) calculations to enable precise monitoring of vegetation in a region (Meneses-Tovar, 2011). Thus, the final choice between raw-color and false-color images depends on the objective of the study. While true-color imagery is ideal for urban mapping and general land cover identification, false-color imagery is widely used in environmental studies, agriculture, and hydrology to enhance subtle differences in land cover that are not visible to the naked eye.

The acquired EO datasets undergo preprocessing steps such as radiometric correction (adjustments for sensor and atmospheric distortions) and geometric correction (aligning of images to a consistent coordinate system). After preprocessing, these datasets can be analyzed and classified using machine learning algorithms for LULC classification (Giri, 2016). LULC classification is typically performed using two main approaches: supervised classification and unsupervised classification.

1. **Supervised Classification**

    - Requires user-defined training data for known land cover types.
    - Machine learning algorithms such as decision trees (DT), random forest (RF), and support vector machines (SVM) learn from training samples to classify the entire image into predefined categories (e.g., forest, urban, water bodies, and agricultural land).

Advantage: Yields higher accuracy when sufficient labeled training data is available (Basanna & Wodeyar, 2013).

2. **Unsupervised Classification**

- Uses algorithms like K-means clustering and (iterative self-organizing data analysis (ISODATA) to group pixels into clusters based on spectral similarities, without requiring prior training data.
- The resulting clusters are later interpreted and labeled by analysts using ground truth data or reference maps.
- Advantage: Useful when no prior land cover information is available which makes it a fast and automated approach (Lang et al., 2008; Waman et al., 2020).

Thus, by integrating ML with EO datasets, enhanced LULC classification of high-resolution LULC of global coverage is provided by agencies like NASA, ESA, JAXA, etc. These organizations provide EO-based land cover products and enable comprehensive temporal and spatial mapping and analysis of land use changes worldwide. Hence, for the LULC analysis of the Narmada River Basin, the ESA WorldCover 10 m v100 dataset (2021) was utilized and processed using GEE. This high-resolution dataset was developed under the ESA's 5th Earth Observation Envelope Programme. It provides a global land cover map for 2020–2021 at a 10-m spatial resolution and is derived from Sentinel-1 and Sentinel-2 imagery. Thus, analysis of the extracted LULC maps of the Narmada River Basin (Fig. 4.8) identified eight distinct LULC classes throughout the river basin. The findings highlight a significant anthropogenic influence, with approximately 57.47% of the basin area dedicated to agriculture. Forested regions are predominantly located in the upper and middle hilly areas and account for 17.01% of the total basin area. Urban built-up areas were primarily concentrated near the river's source and mouth, and they occupied the smallest proportion of the basin (0.95%). However, a more comprehensive assessment of anthropogenic influence requires integrating Earth Observation datasets with other spatial datasets like cartographic and infrastructure maps.

### *4.3.2 Spatial Extent of Human Alterations of Terrestrial Ecosystem of the Narmada River Basin*

One such dataset is the Global Human Modification (gHM) dataset (Kennedy et al., 2019), which provides a detailed measure of anthropogenic alterations to the terrestrial ecosystems. The gHM dataset is developed by The Nature Conservancy and Conservation Science Partners, and it offers a high-resolution (1000 m) global dataset by integrating thirteen distinct human-induced stress parameters which are as follows:

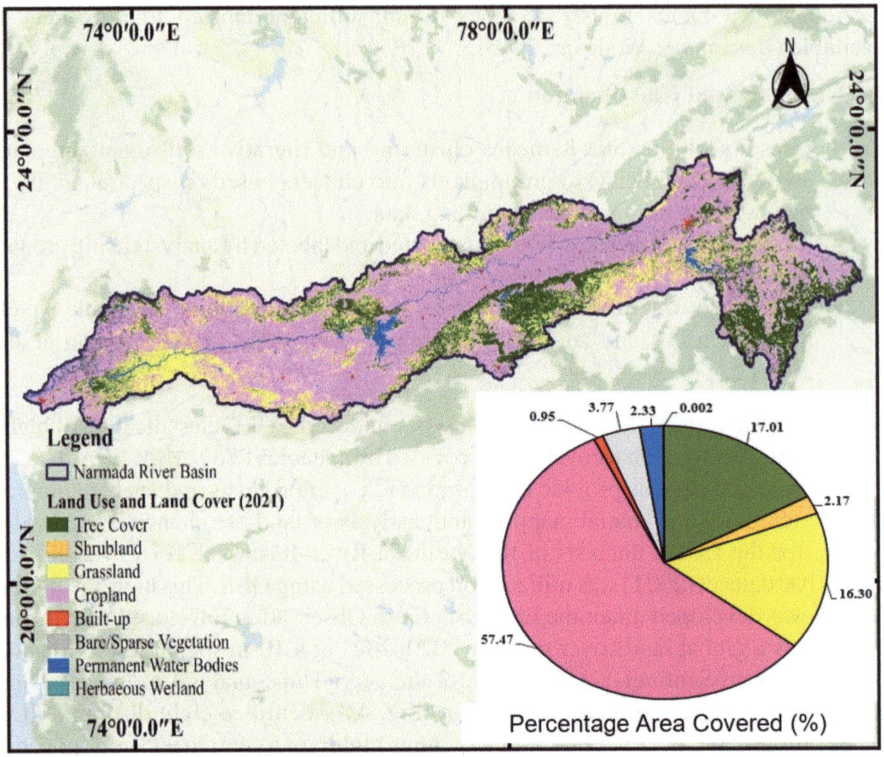


Fig. 4.8 Spatial distribution of LULC in the Narmada River Basin (2021)

- Urban development (population density, urban extent),
- Agriculture (cropland areas, livestock density),
- Transportation (major/minor roads, two-track roads, railroads),
- Infrastructure (power lines, nighttime lights),
- Mining and energy production.

This dataset quantifies human modification levels on a scale of 0 (no modification) to 1 (fully modified) and enables the identification of areas significantly altered by human activity. It is based on the metadata from the year 2016. Following data acquisition, the gHM dataset was extracted and categorized into five levels of human alteration for the Narmada River Basin (Table 4.8).

Further analysis revealed that medium (35.14%) and high (38.21%) levels of human alteration dominated the Narmada River Basin. In contrast, very low alteration was observed in only 0.29% of the area, while very high alteration was limited to 0.21% of the entire basin area (Fig. 4.9). Comparatively, the lower Narmada River Basin, particularly near its mouth, exhibited higher anthropogenic influence, with several zones classified under very high alteration levels. These highly altered regions overlapped with major urban clusters like Jabalpur and Bharuch and hence

**Table 4.8** Classification of terrestrial ecosystem modification levels based on range of alterations

| Sl. No. | Modification level | Range |
|---|---|---|
| 1 | Very low alteration | 0.0–0.2 |
| 2 | Low alteration | 0.2–0.4 |
| 3 | Medium alteration | 0.4–0.6 |
| 4 | High alteration | 0.6–0.8 |
| 5 | Very high alteration | 0.8–1.0 |

further highlight significant human-induced terrestrial ecosystem modifications and their future implications in terms of availability of natural resources and other environmental hazards (Roy, 2024; Roy & Goyal, 2025).

### *4.3.3 Water Stress Under Projected Climate Change Scenario in the Narmada River Basin*

Assessing water stress in the Narmada River Basin requires an integrated approach that combines EO datasets and ML models with advanced hydrological simulations. The recent trends in precipitation variability, intensity, and duration have contributed to seasonal water stress, particularly during the summer months, even in the perennial Narmada River. In this context, the spatial distribution of future water risk can be assessed using the Aqueduct 4.0 dataset (Kuzma et al., 2023). This dataset was developed by the World Resources Institute (WRI). It provides a comprehensive framework that integrates EO-based hydrological modeling with ML-driven risk assessment. This dataset compiles 13 key water risk indicators, categorized into water quantity, water quality, and reputational concerns. Additionally, it incorporates the global hydrological model of PCR-GLOBWB 2, to simulate sub-basin water supply conditions under present and projected future water availability scenarios, based on CMIP6 climate projections (SSP1-RCP2.6, SSP3-RCP7.0, and SSP5-RCP8.5) for 2030, 2050, and 2080 (WRI, 2023). This EO dataset also leverages satellite-derived hydrological variables to refine water availability estimates by incorporating Gravity Recovery and Climate Experiment (GRACE) for groundwater storage and Sentinel-2 for surface water extent. Its integration significantly enhances the spatial accuracy of water stress assessments. Furthermore, ML models in Aqueduct 4.0 facilitate automated classification of high-risk areas, predictive modeling of future water demand, and anomaly detection for early warning systems.

Hence, the Aqueduct Future Annual dataset was accessed and utilized to extract the spatial distribution of projected water stress in terms of water supply, demand, depletion, and variability across the Narmada River Basin. Although the original dataset categorizes water stress levels on a 0–5 scale, for seamless integration with other environmental monitoring datasets, the data were rescaled and normalized to ensure water stress values range from 0–1 across the river basin. Finally, the water

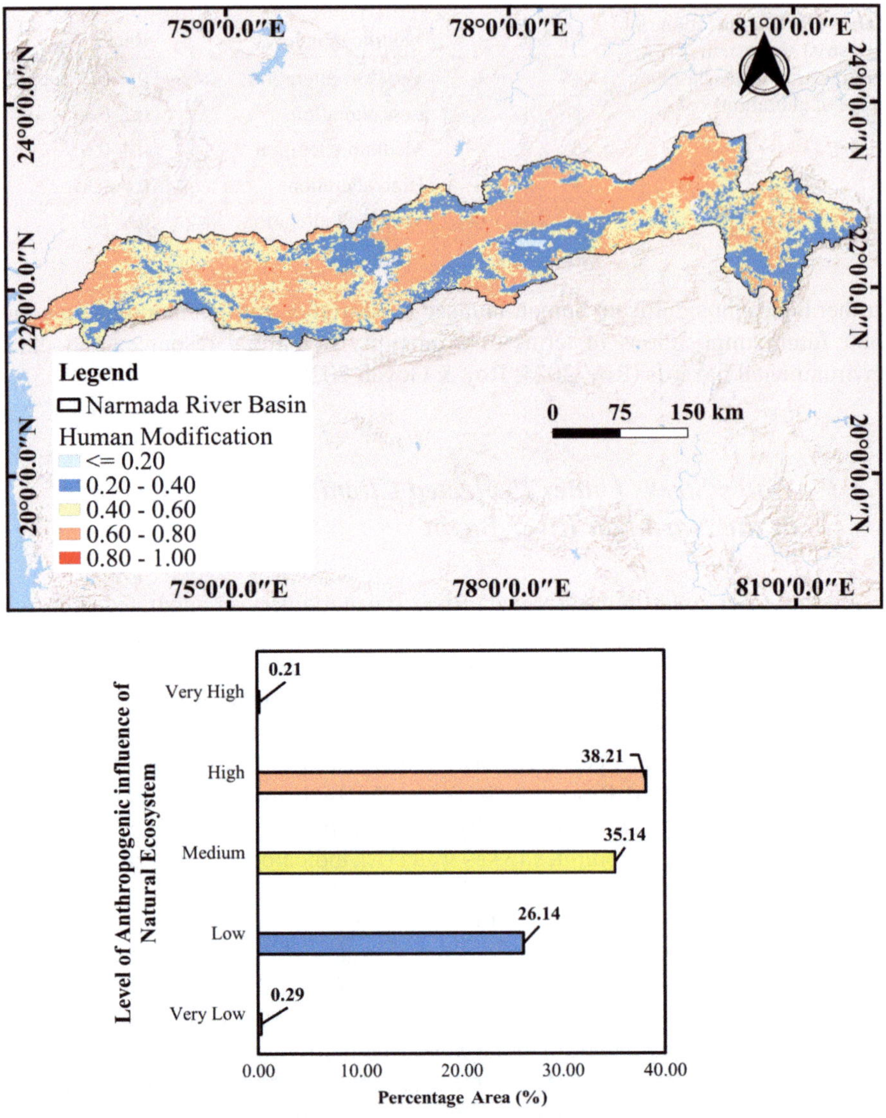


**Fig. 4.9** Spatial distribution of human influence on terrestrial ecosystem (2016)

stress score layer of the Narmada River Basin under the business-as-usual scenario (SSP3-RCP7.0) for the year 2030 was mapped and analyzed (Fig. 4.10). The results reveal that 43.55% of the Narmada River Basin is projected to experience very high-water stress by 2030. Spatially, the lower and middle portions of the basin are particularly vulnerable, where increasing water demand and depletion contribute to

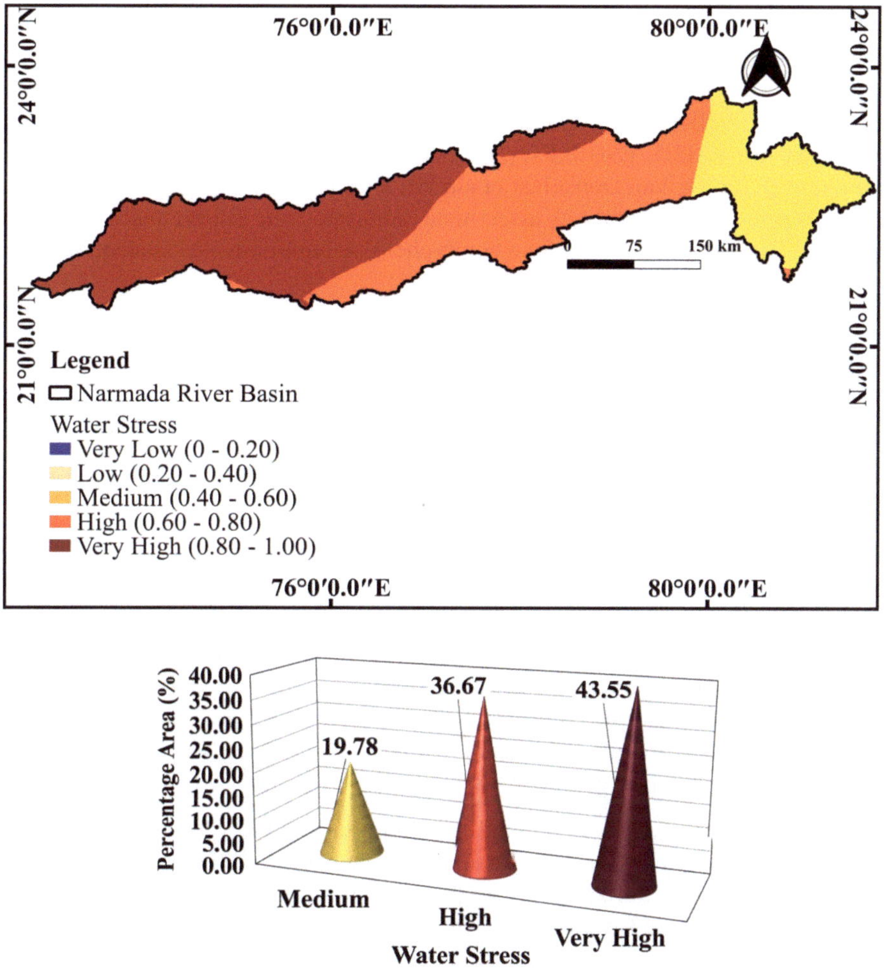


**Fig. 4.10** Spatial distribution of projected water stress levels in the Narmada River Basin under the business-as-usual scenario (2030)

heightened water scarcity. This imbalance between water availability and consumption indicates a significant risk of water stress intensification in the coming decades in the Narmada River Basin. Furthermore, this projection represents a business-as-usual scenario, meaning that if effective water conservation measures and sustainable water management strategies are not implemented, the situation may worsen beyond 2030. Hence, preserving natural water sources, improving groundwater recharge, and enhancing water-use efficiency will be crucial in mitigating future water scarcity across this basin.

Moreover, in addition to human-induced alterations to terrestrial ecosystems and water stress, air quality is a critical environmental factor that must be considered when evaluating a region's overall condition. The availability of Earth Observation datasets like MODIS, VIIRS, and Sentinel-5P enables the remote sensing of air pollution indicators like PM2.5, PM10, $CO_2$, CO, etc., concentrations (Keesari et al., 2021; Kumar et al., 2024; Van Donkelaar et al., 2015). Therefore, an integrated approach that combines hydrological risk assessment, anthropogenic impact analysis, and air quality monitoring is essential for a comprehensive environmental evaluation.

### *4.3.4 Spatial Distribution of Air Quality in the Narmada River Basin*

Air Quality Index (AQI) data of a region can be obtained from several open-source EO datasets like.

- NASA's Land, Atmosphere Near real-time Capability for Earth observation (LANCE): Provides near-real-time data to improve local and national air quality forecasts.
- OpenAQ Platform: Aggregates air quality measurements from over 8,000 locations in 65 countries, offering a comprehensive view of global air quality.
- World Health Organization (WHO) Air Quality Database: Hosts data on air quality for over 600 human settlements in more than 100 countries, serving as a valuable resource for global air quality assessments.

However, in this study, the Global Air Pollution Dataset (2024) which encompasses AQI values for various pollutants across numerous cities worldwide and is hosted on Kaggle was utilized to extract the existing condition and spatial distribution of AQI in the Narmada River Basin for the latest year of 2024. Moreover, to facilitate spatial analysis, the city-level data was interpolated and converted into raster TIFF files with a resolution of 1,000 m. This process enabled the creation of detailed AQI maps, as shown in Fig. 4.11.

The extracted dataset reveals that 94.99% of the Narmada River Basin falls under the moderate air quality category (AQI: 100–200). This is indicative of the fact that while air pollution levels are not severe in this basin, they may cause discomfort for sensitive groups. In 2024, the highest recorded AQI value was 486, whereas the lowest and mean AQI values were 106 and 166, respectively. These values thus further highlight the nonuniformity of spatial distribution of AQI values throughout the river basin. Additionally, 4.59% of the basin experienced poor air quality, which could pose health concerns for vulnerable populations. Moreover, scattered portions of the upper and lower Narmada River Basin were observed to exhibit very poor and severe AQI. Thus, 0.35% of the basin area experienced very poor air quality, while a smaller fraction (0.07%) faced severe pollution levels. The classification system applied in this assessment follows the SAFAR-India Air Quality Index (AQI)

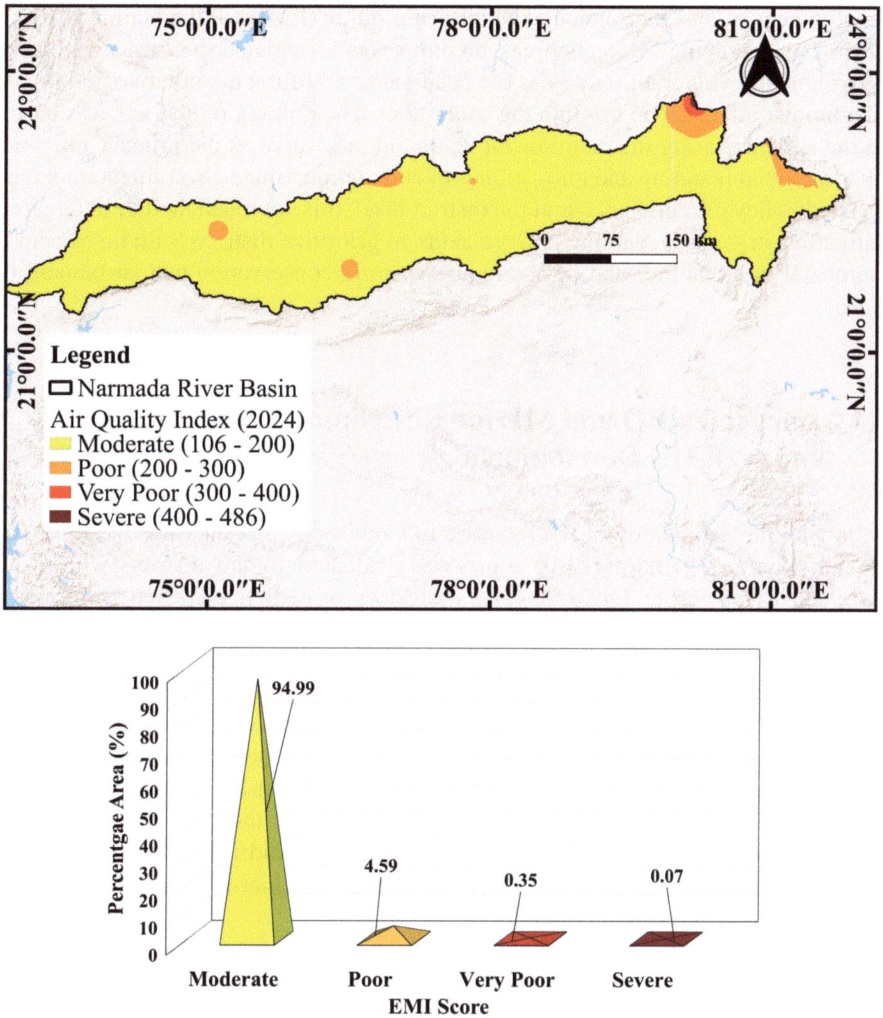


**Fig. 4.11** Spatial distribution of AQI throughout the Narmada River Basin (2024)

system, developed by the Ministry of Earth Sciences, Government of India, for urban air quality monitoring. While the current air quality scenario in the Narmada Basin is not critical, the findings emphasize the importance of proactive mitigation strategies to prevent further environmental degradation. Moreover, in order to comprehensively assess the environmental health of the Narmada River Basin, the AQI spatial distribution layer was normalized to ensure that the values range between 0 and 1, allowing for seamless integration with the other considered environmental parameters. All the three raster datasets were preprocessed and aligned to ensure perfect spatial overlay for further analysis. Thus, these three environmental indicators—human interference, water stress, and air quality—were integrated into the development of an Environmental Monitoring Index (EMI) for the Narmada River Basin. By leveraging ML techniques, an index was formulated to assess the relative environmental vulnerability across the basin, allowing for a quantitative ranking of all administrative districts within the study area. The ranking of districts was based on the EMI, ensuring that administrative boundaries serve as the primary units for environmental planning and mitigation implementation. Since government intervention and policy execution occur at the district level, this approach facilitates targeted mitigation strategies, enabling policymakers to prioritize districts with higher environmental vulnerability and devise region-specific conservation and sustainability plans.

## 4.4 Integrated EO and ML for Environmental Monitoring Index (EMI) Development

Following the acquisition of the required EO environmental datasets, the Environmental Monitoring Index (EMI) score was evaluated for all districts within the Narmada River Basin. Environmental indicators were first extracted from raster datasets using zonal statistics, ensuring that each district was assigned mean values for air quality (AQI), human alteration, and water stress. Subsequently, a random forest regressor (RFR) was trained to model EMI as a function of these three environmental indicators. This ML technique of random forest was selected due to its capability to handle nonlinear relationships between parameters and assess the relative influence of each environmental factor on overall environmental stress. The trained model was then used to predict EMI scores for each district of the river basin. A higher EMI value indicates greater environmental degradation and vulnerability. Based on these EMI scores, the districts were ranked, providing a hierarchical framework for targeted environmental intervention.

Moreover, to enhance interpretability, feature-importance analysis was conducted to identify the most influential environmental variables in EMI computation. The analysis revealed that predicted water stress for 2030 (0.338) played the most significant role in determining EMI values, highlighting it as the dominant environmental pressure shaping regional vulnerability. The remaining two environmental parameters, AQI and human alteration, were observed to have equal influence on the EMI score. The results were visualized through a feature-importance plot to illustrate these contributions (Fig. 4.12b). Finally, the district-wise rankings were categorized into five distinct classes based on EMI score (Fig. 4.12c) using the natural breaks (Jenks) method (Chen et al., 2013). The Jenks optimization algorithm was chosen because it minimizes intra-class variance while maximizing inter-class differences, ensuring that districts with similar environmental conditions are grouped together while maintaining clear distinctions between different levels of environmental stress.

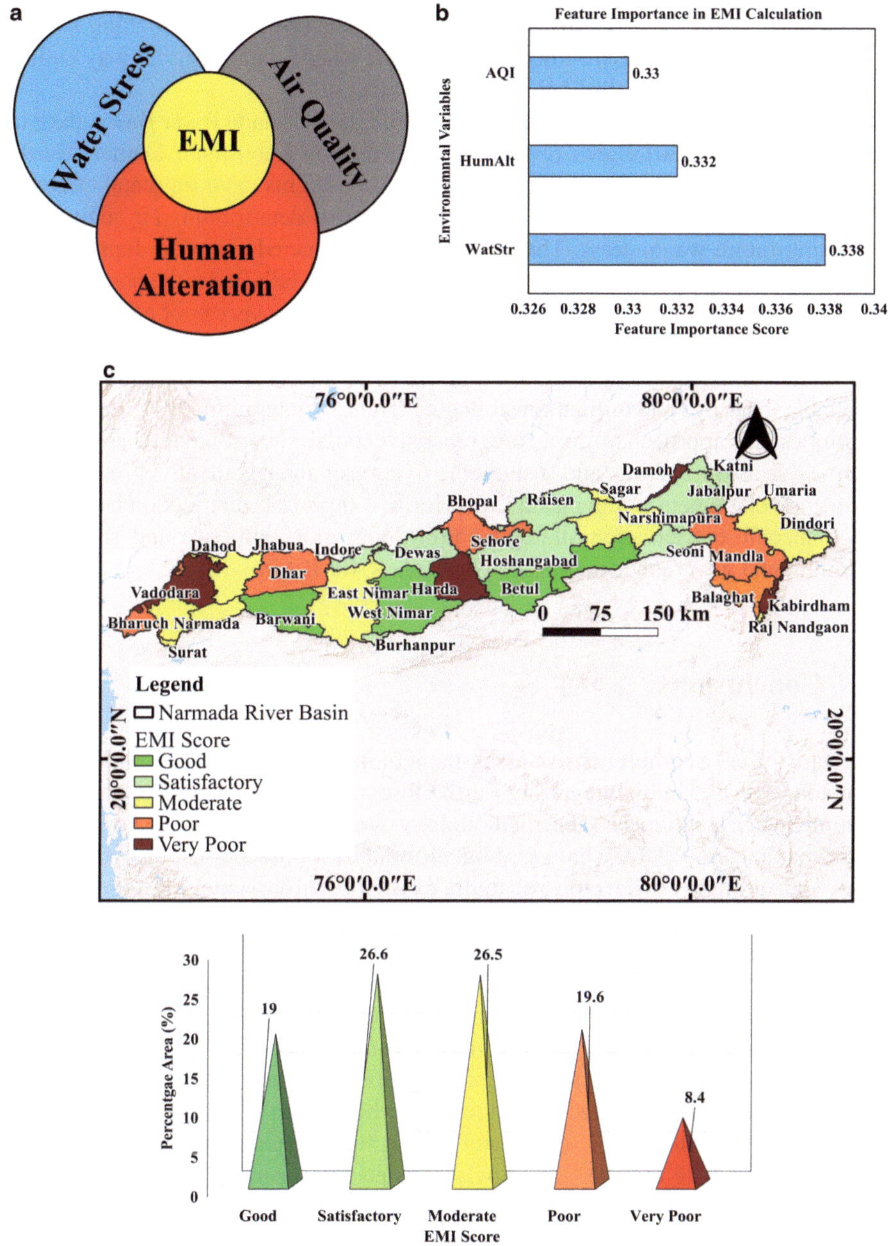


**Fig. 4.12** **a** EMI framework; **b** random forest based feature-importance analysis; **c** EMI-based ranking of the districts of Narmada River Basin

This method is particularly effective for unevenly distributed data, making it well-suited for categorizing environmental risk levels, where some districts may exhibit significantly higher or lower EMI scores than others.

The final results of the district rankings within the Narmada River Basin, derived from the computed EMI scores, reveal that the districts of Kabirdham, Damoh, Harda, and Vadodara exhibit poor environmental conditions. This environmental degradation is primarily driven by intense human intervention, deteriorating air quality, and increasing future water stress. The classification conducted using the Jenks natural breaks method further highlights that 19.6% of the basin falls under the poor environmental status category, while 8.4% is categorized as having very poor environmental conditions based on the selected parameters. This categorization, grounded in current environmental conditions, provides a foundation for the targeted implementation of actionable measures and mitigation strategies. Thus, by integrating EO data with ML techniques, this approach offers a comprehensive tool for environmental assessment, informed decision-making, and sustainable river basin management. Ultimately, this framework empowers policymakers to prioritize high-risk districts and formulate region-specific conservation strategies, promoting more effective and sustainable environmental governance across the Narmada River Basin.

## 4.5 Conclusions

This study offers a comprehensive assessment of the environmental conditions of the Narmada River Basin by integrating Earth Observation (EO) datasets with machine learning (ML) techniques. The methodology focused on analyzing key environmental parameters: LULC changes, human-induced modifications, projected water stress, and air quality to develop a spatially explicit Environmental Monitoring Index (EMI). This index enabled a quantitative ranking of administrative districts, highlighting spatial variations in environmental vulnerability and facilitating targeted policy interventions. Moreover, the EO-based monitoring provided valuable insights into the dynamic landscape of the basin, identifying critical environmental stressors like intensifying human influence, declining air quality, and escalating water scarcity. The classification of land use also revealed significant anthropogenic pressures, with 57.47% of the basin dedicated to agriculture, while urban expansion was primarily concentrated near the river's source and mouth. The assessment of projected water stress under climate change scenarios suggested that nearly 43.55% of the basin is likely to face high to very high-water stress by 2030, underscoring the urgent need for sustainable water resource management. Additionally, air quality analysis indicated moderate to poor AQI levels, with certain districts experiencing severe pollution risks. The integration of EO and ML thus facilitated high-resolution spatial analysis, enabling the development of an EMI framework that ranked districts based on their environmental conditions. The classification of EMI scores using the Jenks natural breaks method also provided a structured categorization of environmental risks, ensuring that districts with similar vulnerabilities were grouped together. The results identified Kabirdham, Damoh, Harda, and Vadodara as the most environmentally vulnerable districts, based on its high human interference, deteriorating air quality, and projected water stress. Moreover, the findings revealed that 19.6% of the basin falls under the poor environmental status category, while 8.4% is classified as very poor, highlighting regions that require immediate intervention.

Conclusively, this study underscores the value of leveraging EO datasets for comprehensive environmental assessments, offering an effective, data-driven approach to integrated environmental monitoring, planning, and decision-making. The EMI framework developed herein serves as a powerful tool for sustainable river basin management, providing policymakers with actionable intelligence to design region-specific mitigation strategies. As environmental pressures continue to escalate, the findings emphasize the urgent need for conservation policies, sustainable land and water management, and proactive climate adaptation measures to preserve the ecological integrity of the Narmada River Basin.

## References


Basanna, R., & Wodeyar, A. K. (2013). Supervised classification for LULC change analysis. *International Journal of Computer Applications, 66*(21), 1–5. https://doi.org/10.5120/11045-6174

Central Water Commission. (n.d.). *Narmada Basin.* India-WRIS. https://indiawris.gov.in/downloads/Narmada%20Basin.pdf

Chang, K. T., & Tsai, B. W. (1991). The effect of DEM resolution on slope and aspect mapping. *Cartography and Geographic Information Systems, 18*(1), 69–77.

Chen, J., Yang, S. T., Li, H. W., Zhang, B., & Lv, J. R. (2013). Research on geographical environment unit division based on the method of natural breaks (Jenks). In *The International Archives of the Photogrammetry, Remote Sensing and Spatial Information Sciences* (Vol. 40), pp. 47–50. https://doi.org/10.5194/isprsarchives-XL-4-W3-47-2013

Dubey, S., Gupta, H., Goyal, M. K., & Joshi, N. (2021). Evaluation of precipitation datasets available on Google Earth Engine over India. *International Journal of Climatology, 41*(10), 4844–4863.

Egorov, A. V., Roy, D. P., Zhang, H. K., Li, Z., Yan, L., & Huang, H. (2019). Landsat 4, 5 and 7 (1982 to 2017) analysis ready data (ARD) observation coverage over the conterminous United States and implications for terrestrial monitoring. *Remote Sensing, 11*(4), 447.

European Space Agency (ESA). (n.d.). *Copernicus DEM: Sentinel-1 derived digital elevation model.* https://spacedata.copernicus.eu/web/cscda/dataset-details?articleId=394198

Fathom Global. (n.d.). *FABDEM: Fine-resolution digital elevation model with vegetation correction.* https://www.fathom.global/fabdem/

Funk, C., Peterson, P., Landsfeld, M., Pedreros, D., Verdin, J., Shukla, S., & Michaelsen, J. (2015). The Climate Hazards Infrared Precipitation with Stations—A new environmental record for monitoring extremes. *Scientific Data, 2*(1), Article 150066.

German Aerospace Center (DLR). (n.d.). *TanDEM-X 90m DEM.* https://tandemx-90m.dlr.de/

Giri, C. P. (Ed.). (2016). *Remote sensing of land use and land cover: Principles and applications.* CRC Press.

Harlin, J. M. (1978). Statistical moments of the hypsometric curve and its density function. *Journal of the International Association for Mathematical Geology, 10*, 59–72.

Hersbach, H., Bell, B., Berrisford, P., Hirahara, S., Horányi, A., Muñoz-Sabater, J., & Thépaut, J. N. (2020). The ERA5 global reanalysis. *Quarterly Journal of the Royal Meteorological Society, 146*(730), 1999–2049.

Horton, R. E. (1945). Erosional development of streams and their drainage basins; hydrophysical approach to quantitative morphology. *Geological Society of America Bulletin, 56*(3), 275–370.

Huffman, G. J., Bolvin, D. T., Braithwaite, D., Hsu, K., Joyce, R., Xie, P., & Yoo, S. H. (2015). NASA Global Precipitation Measurement (GPM) Integrated Multi-satellite Retrievals for GPM (IMERG). *Algorithm Theoretical Basis Document (ATBD)*, Version 4.26.

Indian Space Research Organisation (ISRO). (n.d.). *CARTOSAT-1: High-resolution imaging for topography mapping*. https://www.isro.gov.in/CARTOSAT_1.html

Jain, V., Kumar, S., & Goyal, M. K. (2025). Relationship between daily precipitation extremes and temperature in changing climate across smart cities of Central India. *Journal of Environmental Management, 380*, Article 125036.

Jain, V., Rautela, K. S., & Goyal, M. K. (2023). Ecological restoration: An overview of science and policy regime. *Ecosystem Restoration: Towards Sustainability and Resilient Development*, 1–27.

Japan Aerospace Exploration Agency (JAXA). (n.d.). *ALOS Global Digital Surface Model ALOS World 3D–30m (AW3D30)*. https://www.eorc.jaxa.jp/ALOS/en/aw3d30/index.htm

Jensen, J. R., & Lulla, K. (1987). *Introductory digital image processing: A remote sensing perspective.*

Keesari, T., Goyal, M. K., Gupta, B., Kumar, N., Roy, A., Sinha, U. K., & Goyal, R. K. (2021). Big data and environmental sustainability based integrated framework for isotope hydrology applications in India. *Environmental Technology & Innovation, 24*, Article 101889.

Kennedy, C. M., Oakleaf, J. R., Theobald, D. M., Baruch-Mordo, S., & Kiesecker, J. (2019). Managing the middle: A shift in conservation priorities based on the global human modification gradient. *Global Change Biology, 25*(3), 811–826. https://doi.org/10.1111/gcb.14549

Kumar, R., Goyal, M. K., Surampalli, R. Y., & Zhang, T. C. (2024). River pollution in India: Exploring regulatory and remedial paths. *Clean Technologies and Environmental Policy, 26*(9), 2777–2799.

Kummerow, C., Barnes, W., Kozu, T., Shiue, J., & Simpson, J. (1998). The Tropical Rainfall Measuring Mission (TRMM) sensor package. *Journal of Atmospheric and Oceanic Technology, 15*(3), 809–817.

Kuzma, S., Bierkens, M. F., Lakshman, S., Luo, T., Saccoccia, L., Sutanudjaja, E. H., & Van Beek, R. (2023). *Aqueduct 4.0: Updated decision-relevant global water risk indicators.* World Resources Institute.

Lang, R., Shao, G., Pijanowski, B. C., & Farnsworth, R. L. (2008). Optimizing unsupervised classifications of remotely sensed imagery with a data-assisted labeling approach. *Computers & Geosciences, 34*(12), 1877–1885. https://doi.org/10.1016/j.cageo.2008.02.011

Lehner, B., Verdin, K., & Jarvis, A. (2008). New global hydrography derived from spaceborne elevation data. *Eos, Transactions American Geophysical Union, 89*(10), 93–94.

Lin, C., & Doyog, N. D. (2023). Challenges of retrieving LULC information in rural-forest mosaic landscapes using random forest technique. *Forests, 14*(4), 816.

Melton, M. A. (1957). *An analysis of the relations among elements of climate, surface properties, and geomorphology* (Vol. 11). Department of Geology, Columbia University.

Meneses-Tovar, C. L. (2011). NDVI as indicator of degradation. *Unasylva, 62*(238), 39–46.

Miller, S. L. (1953). A production of amino acids under possible primitive earth conditions. *Science, 117*(3046), 528–529. https://doi.org/10.1126/science.117.3046.528

Narmada Control Authority (NCA). (2025). *Narmada Basin.* https://nca.gov.in/basin.htm

NASA. (n.d.). *ASTER Global Digital Elevation Model (ASTER GDEM).* https://asterweb.jpl.nasa.gov/gdem.asp

NASA. (n.d.). *Shuttle Radar Topography Mission (SRTM).* https://www2.jpl.nasa.gov/srtm/

National Institute of Hydrology (NIH). (1996–1997). *Identification of sampling sites for water quality monitoring in Narmada basin, Madhya Pradesh.* India Water Portal. https://oldiwp.indiawaterportal.org/sites/default/files/iwp2/Identification_of_sampling_sites_for_water_quality_monitoring_in_Narmada_basin__Madhya_Pradesh_NIH_1996_97.pdf

National Oceanic and Atmospheric Administration (NOAA). (n.d.). *ETOPO1 global relief model.* https://www.ngdc.noaa.gov/mgg/global/

Parasuraman, S., Upadhyaya, H., & Balasubramanian, G. (2010). Sardar Sarovar project: The war of attrition. *Economic and Political Weekly, 45*(10), 39–48.

Pérez-Peña, J. V., Azañón, J. M., & Azor, A. (2009). CalHypso: An ArcGIS extension to calculate hypsometric curves and their statistical moments. Applications to drainage basin analysis in SE Spain. *Computers & Geosciences, 35*(6), 1214–1223.

Polar Geospatial Center (PGC). (n.d.). *ArcticDEM: High-resolution digital elevation model for arctic regions.* https://www.pgc.umn.edu/data/arcticdem/

Polar Geospatial Center (PGC). (n.d.). *EarthDEM: High-resolution elevation data for land and ice regions.* https://www.pgc.umn.edu/data/earthdem/

Rakkasagi, S., & Goyal, M. K. (2025). Are rapid onset drying events escalating forest fires across India's ecoregions? *Environmental Research Letters.*

Rakkasagi, S., Edris, S. G., & Goyal, M. K. (2025). A socio-technical framework for assessing flash drought risks to key biodiversity areas: A scientists' warning. *Technology in Society, 82*, Article 102893.

Ray, L. K., & Goel, N. K. (2019). Flood frequency analysis of Narmada River Basin in India under nonstationary condition. *Journal of Hydrologic Engineering, 24*(8), 05019018. https://doi.org/10.1061/(ASCE)HE.1943-5584.0001840

Roy, S. (2024). Decoding the declining mangrove ecosystems: An analysis of carbon dynamics, salinity changes, and deforestation using remote sensing and machine learning. In *AGU24.*

Roy, S., & Chintalacheruvu, M. R. (2023). Assessing long-term changes in the lower Godavari River Basin using geo-spatial technology: An analysis of river bank erosion and accretion. *Sustainable Lifestyle Approaches Towards Environment and Climate Change, EIACP (PC-RP) Special*, 36–49.

Roy, S., & Chintalacheruvu, M. R. (2024a). Enhanced morphometric analysis for soil erosion susceptibility mapping in the Godavari River Basin, India: Leveraging Google Earth Engine and principal component analysis. *ISH Journal of Hydraulic Engineering, 30*(2), 228–244.

Roy, S., & Chintalacheruvu, M. R. (2024b). Google Earth Engine-based morphometric parameter evaluation and comparative analysis of soil erosion susceptibility using statistical and machine learning algorithms in large river basins. *Earth Science Informatics, 17*(1), 75–97.

Roy, S., & Goyal, M. (2025). Climate variability and extremes in Indian River Basins: Trends, regional disparities, and urban risk assessment (No. EGU25-1114). In *Copernicus Meetings.*

Roy, S., Uddin, M. G., Abdelrahman, K., Fnais, M. S., & Abioui, M. (2025). Assessing the impact of digital elevation model resolution on hypsometric analysis in large river basins (India): A non-parametric statistical approach. *Earth Science Informatics, 18*(1), 1–21.

Saksena, S., & Merwade, V. (2015). Incorporating the effect of DEM resolution and accuracy for improved flood inundation mapping. *Journal of Hydrology, 530*, 180–194.

Schumm, S. A. (1963). Sinuosity of alluvial rivers on the Great Plains. *Geological Society of America Bulletin, 74*(9), 1089–1100. https://doi.org/10.1130/0016-7606(1963)74[1089:SOAROT]2.0.CO;2

Scripps Institution of Oceanography (SIO). (n.d.). *SRTM30_PLUS: Global bathymetry and topography.* http://topex.ucsd.edu/WWW_html/srtm30_plus.html

Şener, M., & Arslanoğlu, M. C. (2023). Morphometric analysis in Google Earth Engine: An online interactive web-based application for global-scale analysis. *Environmental Modelling & Software, 162*, Article 105640. https://doi.org/10.1016/j.envsoft.2023.105640

Sharma, S., Dixit, S., Jain, P., Shah, K. W., & Vishwakarma, R. (2008). Statistical evaluation of hydrobiological parameters of Narmada River water at Hoshangabad City, India. *Environmental Monitoring and Assessment, 143*, 195–202.

Shekar, P. R., & Mathew, A. (2022). Erosion susceptibility mapping based on hypsometric analysis using remote sensing and geographical information system techniques. In *River dynamics and flood hazards: Studies on risk and mitigation* (pp. 483–494). Springer Nature Singapore.

Singh, S., & Kumar, A. (2024). Understanding the intricacies of rainfall dynamics using entropy measures. *Journal of Water and Climate Change, 15*(9), 4817–4839.

Singh, S., Kumar, D., Vishwakarma, D. K., Kumar, R., & Kushwaha, N. L. (2024). Seasonal rainfall pattern using coupled neural network-wavelet technique of southern Uttarakhand, India. *Theoretical and Applied Climatology, 155*(6), 5185–5201.

Smith, K. G. (1950). Standards for grading texture of erosional topography. *American Journal of Science, 248*, 655–668.

Strahler, A. N. (1952). Hypsometric (area-altitude) analysis of erosional topography. *Geological Society of America Bulletin, 63*(11), 1117–1142.

Strahler, A. N. (1956). Quantitative slope analysis. *Geological Society of America Bulletin, 67*(5), 571–596.

Strahler, A. N. (1957). Quantitative analysis of watershed geomorphology. *Eos, Transactions American Geophysical Union, 38*(6), 913–920. https://doi.org/10.1029/TR038i006p00913

Technical University of Munich (TUM). (n.d.). *Earth 2014: Global topography and relief model.* https://www.lrg.tum.de/iapg/forschung/topographie/earth2014/

Thakur, T. K., Patel, D. K., Dutta, J., Kumar, A., Kaushik, S., Bijalwan, A., & Ansari, M. J. (2021). Assessment of decadal land use dynamics of upper catchment area of Narmada River, the lifeline of Central India. *Journal of King Saud University-Science, 33*(2), Article 101322.

Toma, M. B., Belete, M. D., & Ulsido, M. D. (2023). Hydrological components and sediment yield response to land use land cover change in the Ajora-Woybo watershed of Omo-Gibe Basin, Ethiopia. *Air, Soil and Water Research, 16*, 11786221221150186. https://doi.org/10.1177/11786221221150186

U.S. Geological Survey (USGS). (n.d.). *GMTED2010: Global multi-resolution terrain elevation data.* https://topotools.cr.usgs.gov/gmted_viewer/viewer.htm

U.S. Geological Survey (USGS). (n.d.). *GTOPO30: Global 30 arc-second elevation model.* https://www.usgs.gov/centers/eros/science/usgs-eros-archive-digital-elevation-global-30-arc-second-elevation-gtopo30

Van Donkelaar, A., Martin, R. V., Spurr, R. J. D., & Burnett, R. T. (2015). High-resolution satellite-derived PM2.5 from optimal estimation and geographically weighted regression over North America. *Environmental Science & Technology, 49*(17), 10482–10491. https://doi.org/10.1021/acs.est.5b02076

Waman, G. R., Mishra, A. D., Deshmukh, R. R., Balakishi, A. I., & Randive, P. U. (2020). LULC-analysis of land-use with the help of unsupervised classification. *Izvestiya Yuzhnogo Federal'nogo Universiteta. Tekhnicheskiye Nauki, (3) (213)*, 184–192.

Wan, Z., Hook, S., & Hulley, G. (2015). MODIS/terra land surface temperature/emissivity 8-Day L3 Global 1 km SIN Grid, Version 6. *NASA EOSDIS Land Processes DAAC*. https://doi.org/10.5067/MODIS/MOD11A2.06

World Resources Institute. (2023). *Aqueduct 4.0: Updated decision-relevant global water risk indicators and future projections.* https://www.wri.org/aqueduct

Zanaga, D., Van De Kerchove, R., Daems, D., De Keersmaecker, W., Brockmann, C., Kirches, G., & Arino, O. (2022). ESA WorldCover 10 m 2021 v200. *European Space Agency (ESA)*. https://doi.org/10.5281/zenodo.5571936

# Chapter 5
# Challenges and Future Directions

## 5.1 Challenges in Integrated Earth Observation and Machine Learning Based Environmental Monitoring

Since EO-ML integrated environmental monitoring holds immense promise for a sustainable future, it is pertinent that the existing major challenges and limitations are effectively addressed. One of the primary issues is data heterogeneity. EO datasets vary in spatial, temporal, and spectral resolutions, and aligning them both with each other and with ML models requires extensive preprocessing. Additionally, this integration demands optimized data harmonization algorithms and domain-specific calibration procedures. The limited availability of high-quality, labeled training datasets particularly in the Global South, further restricts the generalizability and reliability of supervised ML models especially for ecological and biodiversity based environmental monitoring (Andersén, 2010; Gamba, 2012; Roy & Chintalacheruvu, 2024a, 2024b, 2024c; Roy et al., 2024; Rautela et al., 2024a, 2024b; Taylour & Broeders, 2015). Infrastructure-related constraints also pose significant barriers in these countries. It is worth mentioning that cloud-based computational platforms like GEE have alleviated many of these limitations. However, access to localized, validated, and labeled datasets remains scarce and is often not freely available for public utilization. Furthermore, stable Internet connectivity, high-performance computing resources, and sufficient storage capacity continue to challenge institutions and researchers in such resource-constrained regions. Maintenance of EO datasets also is challenging due to the limited space for handling these large EO data archives.

From a technical standpoint, the development, fine-tuning, and validation of ML models followed by the efficient interpretation of their outputs demand specialized expertise. Many of the models, despite being accurate, function as black boxes, thereby making it difficult to interpret or justify the findings. This is especially alarming in high stakes applications like disaster risk assessment or ecological



S. Roy and M. K. Goyal, *Environmental Monitoring with Integrated Earth Observation Data and Machine Learning*, Studies in Computational Intelligence 1278,
https://doi.org/10.1007/978-3-032-19935-5_5

zoning. Moreover, issues like overfitting due to small training samples, data bias, and limited regional transferability further affect these models' robustness and scalability. Finally, ethical and legal concerns regarding data ownership must be addressed with urgency. Finally, questions of data sovereignty and accountability remain underexplored and unresolved as many EO platforms and ML tools rely on open-access datasets and third-party APIs (Doldrina, 2015; Gomes et al., 2020; Rakkasagi et al., 2024, 2025).

## 5.2 The Future of Integrating Earth Observation and Machine Learning for Environmental Monitoring

The future of environmental monitoring lies in the effective integration of EO and ML to enable proactive and globally scalable solutions. With ongoing advancements in satellite missions, the increasing availability of high-resolution, hyperspectral, and multi-temporal EO datasets will significantly enhance the input quality for ML models. This will support more precise and granular monitoring of critical environmental aspects such as crop health, water quality, urban expansion, infrastructure planning, soil degradation, and atmospheric changes (Văduva et al., 2018). Simultaneously, innovations in ML architecture, particularly deep learning models like convolutional neural networks (CNNs) (Chua, 1997), transformers (Amjoud & Amrouch, 2023), and graph neural networks (Wu et al., 2020), etc., are expected to significantly improve the spatial and contextual understanding of EO data for environmental monitoring. These models offer superior performance in pattern recognition, temporal forecasting, and anomaly detection. Furthermore, the integration of physics informed ML (Vadyala et al., 2022) and hybrid models (Didona & Romano, 2015) can reduce dependence on large training datasets while preserving interpretability and scientific reliability.

The democratization and advancements of ML and EO through open platforms (e.g., GEE, NASA's AppEEARS, ESA's SNAP Toolbox) further empowers a broader community of researchers and citizens. This growing accessibility also promotes the evolution of citizen science initiatives for participatory environmental monitoring (Dixit & Pandey, 2024; Lotfian et al., 2021; Roy & Goyal, 2025). Similarly, tools that combine low code interfaces with backend ML functionalities will also make it easier for non-specialists to develop tailored applications without extensive programming skills. Another emerging trend is the adoption of federated learning and edge computing, which enables model training directly on local or remote devices without the need to transfer large datasets. This decentralization enhances data security and enables near-real-time analysis, especially in areas with limited Internet connectivity. Ultimately, the future of EO-ML integration calls for stronger alignment with decision-support systems. Embedding EO-ML outputs into interactive dashboards, early warning systems, and geospatial portals will provide actionable insights to the environmentalists and policymakers for sustainable environmental monitoring.

## 5.3 Bridging the Gap Between Technology and Policy

While technological innovations in EO and ML integrated analysis have revolutionized environmental monitoring, their full potential can only be realized through effective policy integration. Thus, bridging this gap between technical capabilities and decision-making processes is essential for translating analytical insights into meaningful, on-the-ground actions. Despite the availability of high-resolution, real-time environmental data, many regions still struggle to incorporate these insights into planning, regulation, and public communication (Singh & Goyal, 2025). A key barrier lies in the limited transfer of technical literacy between researchers, policymakers, and stakeholders. This disconnect often leads to underutilization or even misinterpretation of analytical outputs. To address this, there is a growing need for user-friendly decision-support systems and interactive dashboards that simplify complex data and render results more actionable for planners, disaster managers, conservationists, and regulators (Ceccato et al., 2014; Dixit et al., 2024, 2025). Policies must also evolve to accommodate the use of EO-ML tools in environmental governance. This includes updating data-sharing regulations, establishing ethical frameworks for AI applications in sustainability monitoring, and ensuring open data access for local authorities and communities.

Integrating EO-ML output into national environmental reporting systems, climate adaptation strategies, and land use planning protocols can significantly enhance accountability, responsiveness, and evidence-based policymaking. The capacity-building efforts through cross-sector training programs, public–private partnerships, and community-based EO-ML initiatives are equally important. Furthermore, collaborations between research institutions, local governments, and international agencies can drive the co-development of tools tailored to policy needs (Alotaibi & Nassif, 2024; Biber, 2011; Dixit et al., 2025; Goyal & Rakkasagi, 2025; Kumar et al., 2024a, 2024b; Lovett et al., 2007; Roy & Chintalacheruvu, 2023; Roy, 2024; Shalu, 2023; Sigh et al., 2024). Ultimately, aligning technological advancements with policy mechanisms improves environmental outcomes along with strengthening transparency, equity, and resilience in governance systems.

## 5.4 Conclusion and Call to Action

Environmental monitoring with integrated EO data and ML has transformed environmental monitoring. This has enabled the evolution of scalable, high-resolution, and automated analysis across diverse environmental scenarios. From detecting long-term precipitation anomalies to mapping mangrove deforestation in climate-sensitive wetland zones, EO-ML frameworks have demonstrated their ability to produce data-driven, actionable insights at both regional and global scales. Thus, these tools have not only improved predictive accuracy and analytical efficiency but have also made environmental intelligence more accessible than ever before.

However, realizing the full potential of this integration requires overcoming persistent challenges in data standardization, computational accessibility, interpretability of models, and cross-sector collaboration. Moreover, as the pace of environmental degradation intensifies, it becomes imperative to bridge technological capabilities with institutional readiness and policy frameworks. The path forward must prioritize inclusivity, transparency, and open innovation. Thus, a collective, cross-disciplinary approach by bringing together data scientists, environmental researchers, local communities, policymakers, and educators is the need of the hour. Similarly, capacity building, ethical governance of AI, and investment in digital infrastructure should be the building block of this transition. In conclusion, integrated EO and ML based environmental monitoring offers a powerful and indispensable toolkit to address the planet's most pressing challenges.

## References


Alotaibi, E., & Nassif, N. (2024). Artificial intelligence in environmental monitoring: In-depth analysis. *Discover Artificial Intelligence, 4*(1), 84. https://doi.org/10.1007/s44162-024-00354-y

Amjoud, A. B., & Amrouch, M. (2023). Object detection using deep learning, CNNs and vision transformers: A review. *IEEE Access, 11*, 35479–35516. https://doi.org/10.1109/ACCESS.2023.3259324

Andersén, J. (2010). A critical examination of the EO-performance relationship. *International Journal of Entrepreneurial Behavior & Research, 16*(4), 309–328. https://doi.org/10.1108/13552551011054507

Biber, E. (2011). The problem of environmental monitoring. *University of Colorado Law Review, 83*, 1.

Ceccato, P., Fernandes, K., Ruiz, D., & Allis, E. (2014). Climate and environmental monitoring for decision making. *Earth Perspectives, 1*, 1–22. https://doi.org/10.1186/2194-6434-1-10

Chua, L. O. (1997). CNN: A vision of complexity. *International Journal of Bifurcation and Chaos, 7*(10), 2219–2425. https://doi.org/10.1142/S0218127497001807

Didona, D., & Romano, P. (2015, January). Hybrid machine learning/analytical models for performance prediction: A tutorial. In *Proceedings of the 6th ACM/SPEC International Conference on Performance Engineering* (pp. 341–344). https://doi.org/10.1145/2668930.2695534

Dixit, S., & Pandey, K. K. (2024). Spatiotemporal variability identification and analysis for non-stationary climatic trends for a tropical river basin of India. *Journal of Environmental Management, 365*, Article 121692.

Dixit, S., Pandey, K. K., & Kumar, S. (2025). Enhancing non-stationary analysis of extreme precipitation through a precise extreme event extraction approach (No. EGU25-264). In *Copernicus Meetings*.

Dixit, S., Pandey, K. K., & Shukla, D. (2024). Decadal hydrological impact assessment of evolving land use and land cover in an Indian river basin: A multi-model approach. *Journal of Water and Climate Change, 15*(9), 4418–4433.

Doldrina, C. (2015). Open data and earth observations: The case of opening up access to and use of earth observation data through the global earth observation system of systems. *Journal of Intellectual Property, Information Technology and Electronic Commerce Law, 6*, 73.

Gamba, P. (2012). Human settlements: A global challenge for EO data processing and interpretation. *Proceedings of the IEEE, 101*(3), 570–581. https://doi.org/10.1109/JPROC.2012.2227030

Gomes, V. C., Queiroz, G. R., & Ferreira, K. R. (2020). An overview of platforms for big earth observation data management and analysis. *Remote Sensing, 12*(8), 1253. https://doi.org/10.3390/rs12081253

Goyal, M. K., & Rakkasagi, S. (2025). *Technology–policy interface for combating forest loss and fires*. Springer Nature.

Kumar, R., Mishra, A., & Goyal, M. K. (2024a). Wastewater management policies: a perspective and roadmap for India. *Current Science (00113891)*, (7).

Kumar, R., Mishra, A., & Goyal, M. K. (2024b). Water neutrality: concept, challenges, policies, and recommendations. *Groundwater for Sustainable Development*, 101306.

Lotfian, M., Ingensand, J., & Brovelli, M. A. (2021). The partnership of citizen science and machine learning: Benefits, risks, and future challenges for engagement, data collection, and data quality. *Sustainability, 13*(14), 8087. https://doi.org/10.3390/su13148087

Lovett, G. M., Burns, D. A., Driscoll, C. T., Jenkins, J. C., Mitchell, M. J., Rustad, L., & Haeuber, R. (2007). Who needs environmental monitoring? *Frontiers in Ecology and the Environment, 5*(5), 253–260. https://doi.org/10.1890/1540-9295(2007)5[253:WNEM]2.0.CO;2

Rakkasagi, S., Edris, S. G., & Goyal, M. K. (2025). A socio-technical framework for assessing flash drought risks to key biodiversity areas: A scientists' warning. *Technology in Society, 82*, Article 102893.

Rakkasagi, S., Goyal, M. K., & Jha, S. (2024). Evaluating the future risk of coastal Ramsar wetlands in India to extreme rainfalls using fuzzy logic. *Journal of Hydrology, 632*, Article 130869.

Rautela, K. S., Singh, S., & Goyal, M. K. (2024a). Characterizing the spatio-temporal distribution, detection, and prediction of aerosol atmospheric rivers on a global scale. *Journal of Environmental Management, 351*, Article 119675.

Rautela, K. S., Singh, S., & Goyal, M. K. (2024b). Resilience to air pollution: A novel approach for detecting and predicting aerosol atmospheric rivers within earth system boundaries. *Earth Systems and Environment*, 1–23.

Roy, S. (2024). Decoding the declining mangrove ecosystems: An analysis of carbon dynamics, salinity changes, and deforestation using remote sensing and machine learning. In *AGU24*.

Roy, S., & Chintalacheruvu, M. R. (2023). Assessing long-term changes in the lower Godavari River Basin using geo-spatial technology: An analysis of river bank erosion and accretion. *Sustainable Lifestyle Approaches Towards Environment and Climate Change, EIACP (PC-RP) Special*, 36–49.

Roy, S., & Chintalacheruvu, M. R. (2024a). Google Earth Engine-based morphometric parameter evaluation and comparative analysis of soil erosion susceptibility using statistical and machine learning algorithms in large river basins. *Earth Science Informatics, 17*(1), 75–97.

Roy, S., & Chintalacheruvu, M. R. (2024b). Delineating hydro-geologically constrained groundwater zones in the Himalayan River basins of India through an innovative ensemble of hypsometric analysis and machine learning algorithms. *Earth Science Informatics, 17*(1), 501–526.

Roy, S., & Chintalacheruvu, M. R. (2024c). Enhanced morphometric analysis for soil erosion susceptibility mapping in the Godavari River basin, India: Leveraging Google Earth Engine and principal component analysis. *ISH Journal of Hydraulic Engineering, 30*(2), 228–244.

Roy, S., & Goyal, M. (2025). Climate variability and extremes in Indian River Basins: Trends, regional disparities, and urban risk assessment (No. EGU25-1114). In *Copernicus Meetings*.

Roy, S., Rao, C. M., & Abioui, M. (2024). Evaluation of non-stationary spatial relationship between meteorological-environmental parameters and PM2.5. *Advances in Space Research, 73*(8), 4106–4124.

Shalu, S. G. (2023). Environmental monitoring with machine learning. *EPRA International Journal of Multidisciplinary Research (IJMR), 9*(6), 208–212.

Singh, S., & Goyal, M. K. (2025). A review of India's water policy and implementation toward a sustainable future. *Journal of Water and Climate Change, 16*(2), 493–510. https://doi.org/10.2166/wcc.2023.285

MIX
Papier aus verantwortungsvollen Quellen
Paper from responsible sources
FSC® C105338

If you have any concerns about our products,
you can contact us on
**ProductSafety@springernature.com**

In case Publisher is established outside the EU,
the EU authorized representative is:
**Springer Nature Customer Service Center GmbH**
**Europaplatz 3, 69115 Heidelberg, Germany**

Printed by Libri Plureos GmbH
in Hamburg, Germany